CRISTHIANE Cattani

Microbial decontamination methods for pig carcasses

CRISTHIANE Cattani

Microbial decontamination methods for pig carcasses

Microbial analyses using alternative methods

ScienciaScripts

Imprint

Any brand names and product names mentioned in this book are subject to trademark, brand or patent protection and are trademarks or registered trademarks of their respective holders. The use of brand names, product names, common names, trade names, product descriptions etc. even without a particular marking in this work is in no way to be construed to mean that such names may be regarded as unrestricted in respect of trademark and brand protection legislation and could thus be used by anyone.

Cover image: www.ingimage.com

This book is a translation from the original published under ISBN 978-613-9-66612-6.

Publisher:
Sciencia Scripts
is a trademark of
Dodo Books Indian Ocean Ltd. and OmniScriptum S.R.L publishing group

120 High Road, East Finchley, London, N2 9ED, United Kingdom
Str. Armeneasca 28/1, office 1, Chisinau MD-2012, Republic of Moldova, Europe
Printed at: see last page
ISBN: 978-620-8-04288-2

I dedicate this work to my family and especially to my grandparents Sebastião (in Memorian) and Gonçala de Oliveira.

ACKNOWLEDGEMENTS

To the Federal University of Santa Catarina and the Postgraduate Programme in Food Science, for the opportunity to carry out this work.

To the Ministry of Agriculture, Livestock and Food Supply for making this stage of my professional training possible and supporting it, especially to my friends at the Federal Superintendence of Santa Catarina, Renato Gerszevski and Francisco A. Powell Van Castelee. Powell Van de Castelee.

To my friends and work colleagues: Antônio Rotta, Leonardo Mulitemo and Marcelo Teixeira, you were fundamental to the execution of this project.

To Cooperativa Central Oeste Catarinense, for the important working partnership, especially to Rodicler Bortolucci and his entire team, for their invaluable help at every stage of the work.

The Santa Catarina State Research and Innovation Support Foundation (FAPESC) and PURAC do Brasil for their financial support for this project.

A special thank you to my supervisor, Professor Dr Cleide Rosana Wemeck Vieira, for accepting the supervision of this work, for her teachings, for her friendship and for welcoming me to the Postgraduate Programme in Food Science.

To my friends in the food microbiology laboratory at UFSC, for all the experiences we've had together over the last few years. The moments we spent together will never be forgotten.

To Professor Dr Paulo José Ogliari for his patience and help with the statistical analyses.

To Sérgio Souza for his friendship and bureaucratic support.

To all the staff at NUMICAL, especially my friend and competent professional Pedro Ivo Pinheiro Fucks, who helped me with all the analyses without any effort.

To my friends and family, especially my husband Ari and my son Guilherme, who have always been there for me throughout this long journey.

To God and life.

SUMMARY

The aim of this study was to evaluate the application of hot water between 76°C and 78°C and 1.5% lactic acid on the surface of pork carcasses with a view to microbial reduction. Four treatments were used, T1, T2, T3 and T4 with water at different temperatures followed or not by sprinkling of lactic acid. Alternative and conventional microbiological diagnostic methods for detecting microorganisms were used and compared. Swabs were taken from naturally contaminated pig carcasses during the normal slaughter process in a slaughterhouse under Federal Inspection in the state of Santa Catarina, and the microorganisms investigated were aerobic mesophiles, *enterobacteria, Escherichia coli, Salmonella* spp., *E. coli* 0157 by conventional methodology, and alternatives by PETRIFILM™ and TEMPO® . There were no positive results for *E. coli* 0157. *Salmonella* spp. was detected in 5 (6.58%) of the 76 carcasses sampled, and the serotypes found were Thyphimurium, Agona and Derby. The treatment that showed the best reduction in the number of contaminated carcasses was the treatment that consisted of bathing the carcasses in water with a temperature between 76°C and 78°C followed by spraying them with lactic acid at a concentration of 1.5%. Pearson's correlation coefficient showed positive correlations (0.67 to 0.83) between the methodologies used for counting enterobacteria and mesophilic microorganisms, but did not show a good correlation for diagnosing *E. coli* (0.39). The use of the alternative methods tested in place of the conventional methodology can be recommended for diagnosing enterobacteria and mesophilic microorganisms because of the agreement between the results found, plus the speed of these methodologies.

Keywords: Lactic acid. Hot water. Pork carcasses. Decontamination.

SUMMARY

CHAPTER 1

INTRODUCTION

Due to the sanitary conditions of animals destined for consumption and the demands caused by the Second World War, which broke out in Europe in the middle of the 20th century, the first slaughterhouses were set up in Brazil with the creation of the Veterinary Services in 1915. These services were responsible for the sanitary inspection of animals and their regulation, with the first rules on the *ante* and *post mortem* inspection of slaughter animals and the hygiene of the process (PARDI, 1996).

With the aim of increasing microbiological safety and hygiene in the slaughter process, from 1997 onwards, slaughterhouses operating under the rules of the Federal Inspection Service in Brazil for the production of meat and animal products had to implement Good Manufacturing Practice (GMP) programmes and Hazard Analysis and Critical Control Point (HACCP) programmes (BRASIL, 1997). However, even with the implementation of these programmes in slaughterhouses, factors such as the health *status of* the herd, the transport of the animals from the farm to the slaughterhouse, the resting time in the slaughterhouse, fasting and the water diet of the batch before slaughter can affect the microbiological condition of pig carcasses (HUIS in't VELD; MULDER; SNIJDERS, 1994).

There are many hazards that can be detected in the *ante-mortem* inspection of pigs and that can pose a risk to public health. However, current hazards, such as pathogens that cause food-borne infections, are not detected by traditional visual inspection at *ante-mortem* or *post-mortem* (SCOTT et al., 2008).

Microorganisms belonging to the *Enterobacteriaceae* family, including *Escherichia coli* and its pathogenic serotypes - such as 0157 - along with *Salmonella* spp. and other enterobacteria, are some of the microbiological hazards that must be controlled in fresh meat (SOFOS, 2008; LARA et al., 2011).

Pork, like other meats from different animal species, can carry these pathogens from the farm to the consumer's table. The presence of these micro-organisms in meat is the result of contamination of live animals, equipment, handlers and the environment. Control measures must be adopted to prevent the spread of pathogens from slaughtered animals to humans (NESBAKKEN; SKJERVE, 1996; SOFOS; BELK; SMITH, 1999).

Therefore, bacterial decontamination methods may be necessary as an effective intervention to reduce contamination after slaughter (SWANENBURG et al., 2001). These carcass decontamination processes are based on immersion, washing or spraying with water or chemical solutions and are used in several countries, such as the United States of America (USA), Canada and Australia. They can be designated as a critical control point in the HACCP programmes developed by industries (SOFOS; BELK; SMITH,1999). These techniques are not used in Brazil, as their use is not provided for in the legislation.

To assess the microbiological condition of these carcasses, conventional or so-called traditional laboratory methods are used to carry out microbiological tests that detect and count viable bacterial cells in food. These methods are sensitive, inexpensive and can provide qualitative and quantitative information on the microorganisms present in the sample, but they take several days to obtain results, as they depend on the ability of the microorganisms to multiply so that there are visible colonies (BOER; BEUMER, 1999). Similarly, rapid laboratory methods need to be included in the diagnostic routine to provide adequate information on the possible presence of pathogens in raw materials and finished products, to control the production process and to monitor hygiene and cleaning practices (BOER; BEUMER, 1999).

In this context, it is important to minimise the health risks present from primary production to the slaughter of these animals and to use rapid microbial control methods during the process.

1.1   OBJECTIVES

The objectives of this study were divided into general objectives:

To evaluate the influence of the use of lactic acid and hot water in the washing of pig carcasses on the microbial population, under normal slaughter conditions, using conventional laboratory methods and alternative methods.

And specific objectives:

a) to evaluate the microbiota present on the surface of pork carcasses by means of the total count of mesophilic aerobic microorganisms, enterobacteria, *E. coli*, *Salmonella* spp. and *E. coli* 0157 using different conventional and alternative methodologies;

b) to evaluate the reduction in contamination of pork carcasses with the use of decontamination processes: 1.5% lactic acid and water at temperatures between 76°C and 78°C;

c) to compare the results of the microorganism counts found by the conventional methodology with the results of the alternative Petrifilm ™ and TEMPO® methodologies;

d) search for *Salmonella* spp. by conventional methodology.

e) search for *E. coli* 0157 using the VIDAS® System.

CHAPTER 2

LITERATURE REVIEW

To support this research, the study was organised into eight sections: Brazilian pork; primary pig production; the pig slaughter process; pig microbiota; indicator microorganisms; pathogenic microorganisms; carcass decontamination methods; and microbiological detection methods.

2.1 BRAZILIAN PORK

Today, the state of Santa Catarina, in the south of Brazil, stands out in terms of pig slaughter and is considered by the International Organisation of Epizootics (OIE) to be free of foot-and-mouth disease without vaccination. The state is responsible for slaughtering 7.8 million pigs in 25 slaughterhouses under the Federal Inspection Service, representing 27.40 per cent of the national pig slaughter, which was 28.5 million in 2010. Pork is the most consumed protein in the world, with a production of 115 million tonnes, almost half of which is produced in China and another third in the European Union (EU) and the United States of America (USA).

Brazil's share of the world market has grown in importance. The country is the fourth largest producer, with 3% of production and 11% of exports. The international pork trade moves 5.4 million tonnes and generates annual revenues of approximately 11.9 billion dollars. It is concentrated in five importing countries (Japan, the Russian Federation, Mexico, South Korea and Hong Kong).

The United States, the European Union, Canada, Brazil and China account for 96 per cent of world exports.

The main highlight of recent years is the performance of Brazilian foreign sales, which in ten years have increased their share of world exports from 4% to 11% (Figure 1). Despite sanitary barriers, increased European subsidies and growing international competition, Brazilian exports have grown above the average of its competitors (BRAZILIAN ASSOCIATION OF THE PIGMEAT PRODUCING AND EXPORTING INDUSTRY, 2010).

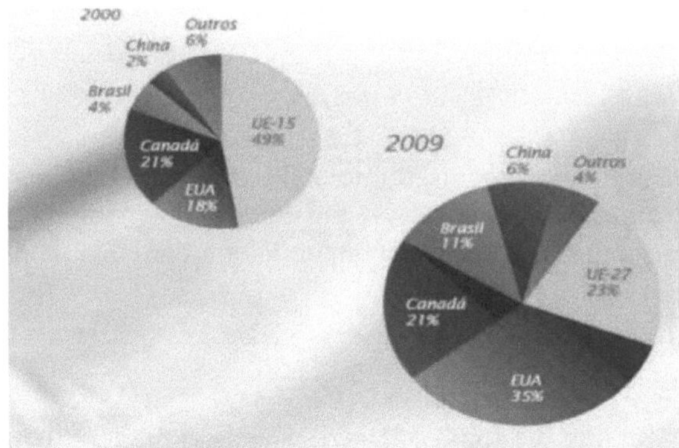

Figure 1 - Brazil's share of the world pork export market in 2000 and 2009

Source: Brazilian Pork Producing and Exporting Industry Association (2010)

2.1 PRIMARY PIG PRODUCTION

The domestic pig - *Sus scrofa domesticus* - belongs to the class of Mammals, order of Ungulates, suborder of Artidactyls, family of Suidae. It forms the genus Sus *(Sus scrofa, Sus vitatus)* (VIANNA, 1974). It was domesticated around 2500 BC and became considerably important in the Greco-Roman period for its meat, when legs were salted and smoked and sausages were produced (LAWRIE, 2005). In Brazil, pig farming can be divided into industrial (technified) and subsistence, with the presence of family, employer and business producers. Its productivity has grown steadily since 2004, with technological advances in genetics, health, nutrition, facilities, management and animal welfare, resulting in an increase in technical efficiency (feed conversion and sow productivity) and the quality of the animals delivered for slaughter (lean meat yield of the carcass), with the southern region of Brazil being the most important area for pig production in the country (MIELE; MACHADO, 2010).

Primary production should receive special attention through pathogen control and surveillance programmes, which increase the risk of food-borne infections in humans (CORTINAS et al., 2009). The spread of infections between animals can be substantially reduced with the development of

strategic interventions in herds, contributing to the protection of the food chain, from the farm to the consumer's fork (BEARSON; BEARSON, 2011). However, the inspection of carcasses and hygiene in slaughterhouses and other handling processes alone are not enough to guarantee the safety and sanitary quality expected (BLAHA, 2001).

Meat hygiene is defined as "all the conditions and measures necessary to guarantee the safety and suitability of meat at all stages of the food chain" (CODE..., 2004a). In this context of meat hygiene, we should apply appropriate measures to protect public health and achieve satisfaction of the quantitative results criteria required to control hazards (CONTROL..., 2006).

Reducing the levels of contaminated carcasses at the slaughterhouse can be achieved by identifying and controlling the sources of contamination at all stages of production, especially before the pre-slaughter period (BESSA; COSTA; CARDOSO, 2004; JORDAN et al., 2006; LO FO WONG et al., 2002; SCHWARZ et al., 2009). Therefore, the ability to control diseases caused by micro-organisms present in meat and meat products will be effectively achieved by reducing the prevalence of infection in animals still on farms, as well as by intervening at other stages of transmission from animal to man (HUMPHREY; JORGENSEN, 2006).

2.2   PIG SLAUGHTERING PROCESS

This section covers all the stages of slaughter in a pig slaughterhouse, from the arrival of the animals to the dispatch of the carcass.

2.2.1 Pre-slaughter handling

Pre-slaughter handling begins when the pigs are still on the farm, when they are loaded, transported by lorry and unloaded at the slaughterhouse, where they are grouped into batches and kept in the pens for between 6 and 24 hours (BRASIL, 1952; BRASIL, 1995). These procedures must comply with animal welfare precepts in order to minimise the risk of falls, fights and pre-slaughter stress, which can compromise the health quality of the carcass during and after slaughter (DALLACOSTA et al, 2007; GREGORY, 2008).

During this time, when the pigs are waiting in the slaughterhouse pen, they are fasted and fed a water diet in order to empty their intestinal contents (BRASIL, 1952; BRASIL, 1995). These animals can be asymptomatic carriers of pathogenic bacteria and serve as a source for meat contamination (LARA et al., 2011; NESBAKKEN et al., 2003; SOFOS; BELK; SMITH, 1999) from the digestive tract, instruments and equipment involved in the process, with cross-contamination occurring from both sources (RIVAS; VIZCAINO; HERRERA, 2000).

The slaughterhouse pens (Figure 2) become sources of contamination between the different batches of pigs in the pre-slaughter period, as they receive faecal material from animals of different health levels, and various enterobacteria such as *Salmonella* sp. can be found in the pigs *or on the floor, Yersinia enterocolitica and Escherichia coli can be found* in the pigs or on the floor of the pen (BINTER et al, 2011; FRANSEN et al, 1996; LO FO WONG et al, 2002; SWANENBURG et al, 2001; VIEIRA PINTO et al, 2010).

Figure 2 - Sties at the slaughterhouse where the pigs are kept on a <u>and water diet before</u> slaughter
Source: Rotta (2010)

Bacteria enter slaughterhouses through animals (hair, skin, feet); inside animals, asymptomatic carriers (gastrointestinal tract) (BAGGESEN et al, 1996) and also through people. There is no specific and direct visual inspection procedure for detecting some pathogenic bacteria, such as

Salmonella spp., which can be transmitted to consumers (SAIDE-ALBORNOZ et al., 1995) through the meat of these animals during slaughter and processing (BEARSON; BEARSON, 2011). *Ante-mortem* inspection of pigs is carried out pre-slaughter, when the animals are in the slaughterhouse pen, and removes excessively dirty and obviously sick animals from slaughter, but it alone cannot prevent the slaughter of pigs with potential pathogens in the gut or on the skin. The *Codex Alimentarius* defines *ante-mortem* inspection as "procedures and tests" to be carried out by the competent authority, based on scientific knowledge, to achieve an integrated public and animal health objective (CODE..., 2009). This procedure is carried out routinely in all slaughterhouses, from the moment the animals are unloaded until they enter the slaughter area (BRASIL, 1995).

2.2.2 Slaughter

According to Brazilian legislation (BRASIL, 1995), the slaughter area is used for operations from stunning to the entry of the carcasses into the cooling chambers, including the space for the final inspection of the carcasses, called the Final Inspection Department (DIF). The final inspection of the carcasses is a *post-mortem* inspection procedure in which the inspector judges whether or not the carcass is fit for human consumption. Slaughter operations (Figure 3) are carried out in two distinct and physically separate areas, called the dirty area and the clean area, where the following operations are carried out:

Dirty area: stunning, bleeding, washing after bleeding, scalding, shaving, scalding and toileting. Polishing is optional and toileting includes removing the remaining hooves, hair and middle ear;

Clean area: occlusion of the rectum, abdominal and thoracic opening, opening of the jowl, inspection of the head and jowl, removal and inspection of organs, longitudinal division of the carcasses, inspection of the carcasses, removal of the tail, head, rump and legs, toileting of the carcass and final washing.

During slaughter, large quantities of carcasses from different origins are handled at the same time and in very close quarters, and there may be several pathogens present, coming from animals from different herds. This creates an opportunity for cross-contamination or the spread of these pathogens in the environment, on equipment or on carcasses (LO FO WONG et al., 2002).

Emerging pathogens such as *Listeria monocytogenes, Escherichia coli* O157:H7 (SOFOS et al., 1999) and *Salmonella* sp. are responsible for increasing the risk of carcass contamination at slaughter (DE BUSSER et al., 2011; KICH et al., 2005).

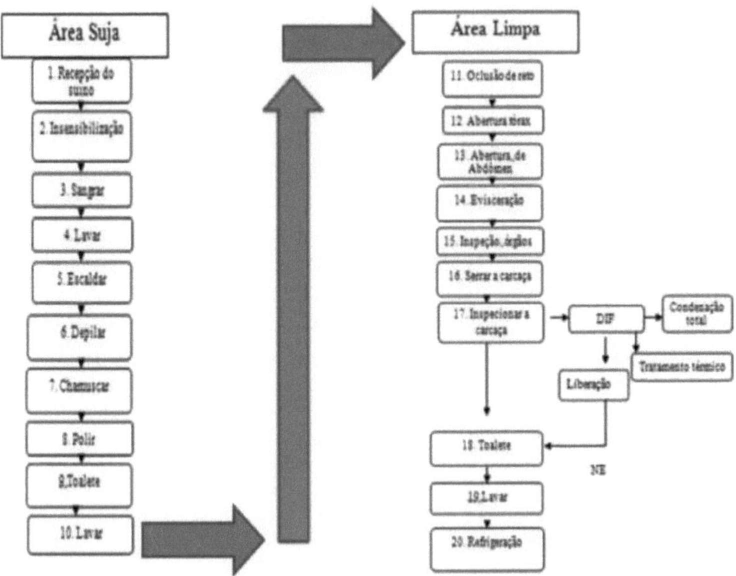

Figure 3 - Flowchart of pig slaughter procedures

Source: Adapted from Ordinance 711/95 (1995)

2.2. 3Dirty area operations

The pig can be stunned by electronarcosis, electrocution or carbon dioxide and the animal must lose consciousness (GREGORY, 2008). The animal is then bled using sanitised knives to make incisions in the large vessels, which causes the blood to drain rapidly, profusely and completely, leading to hypovolemic shock and the animal's death (ANIL; WHITTINGTON; MCKINSTRY, 2000).

The carcasses are then hung by the rear limb and follow overhead rails to the shower, where the carcass is washed only with water at room temperature. After washing, the carcass is scalded and shaved using equipment designed for this purpose, with water controlled at a minimum temperature of 62°C (BRASIL, 1995). This stage reduces the microbiological counts on the surface of the carcass (BOLTON; PEARCE; SHERIDAN, 2002; INTERNATIONAL COMMISSION ON MICROBIOLOGICAL SPECIFICATIONS FOR FOODS, 2005; LO FO WONG et al, 2002) and the risk of contamination by *Salmonella* spp. via the scalding water can be minimised if the temperature of the scalding water is above 60°C and if the pig remains in the scalding tank for at least 1 to 4 minutes (BOLTON et al., 2003).

The carcasses are then polished to remove hair residue. At this stage, there may be an increase in contamination caused by the contact between the polishing whips and the carcass. This contamination is eliminated in the next stage of the slaughter process, which is scalding (BOLTON; PEARCE; SHERIDAN, 2002; BERENDS et al., 1997).

In the scorching or flaming stage, the carcasses pass through equipment with nozzles that shoot flames onto their surface, destroying the superficial micro-organisms on the pig's skin and burning the hair residue (INTERNATIONAL COMMISSION ON MICROBIOLOGICAL SPECIFICATIONS FOR FOODS, 2005).

Flaming or scorching is referred to as destroying microorganisms by the direct action of heat (Figure 4). The low percentage of *Salmonella* spp. found after *swabbing* pig carcasses sampled after flambéing at slaughter is achieved by a combination of processing steps (SAIDE-ALBORNOZ et al., 1995). This operation is followed by toileting, which can be a point of recontamination of the carcass, and subsequent washing of the carcasses with water at room temperature, bringing to an end the operations of the dirty slaughter zone (BRASIL, 1995). All the water used in the industry must be chlorinated and have a maximum free residual chlorine content at any point in the supply system of 2 mg/L, which is the drinking water standard for human consumption (BRASIL, 2011).

Figure 4 - Buckling operation in the dirty pig slaughter area
Source: Cattani (2010)

2.2.   4Clean area operations

The first operation in the clean area is occlusion of the rectum, carried out manually or mechanically by separating the final portion of the rectum from the adjacent tissues, bagging it and tying it (BRASIL, 1995). The aim of this procedure is to prevent the faecal content from leaving the rectum and coming into contact with the carcass, and it is extremely important for minimising faecal contamination (BERENDS et al., 1997; PEARCE et al., 2004). Next, the abdominal and thoracic cavities are opened, consisting of the ventral median cut of the respective walls, followed by exposure of the internal organs to the outside of the sectioned cavities. This procedure is called evisceration and is considered an important control point for faecal contamination in carcasses (BOLTON; PEARCE; SHERIDAN, 2002; DE BUSSER et al., 2011).

2.2.4.1 *Post-mortem*   inspection procedures

Post-mortem inspection consists of examining the carcass, parts of the carcass, cavities, organs, tissues and lymph nodes by visualisation, palpation, olfaction and incision when necessary (Figure 5). This stage is divided into inspection lines, which include inspection of the head, jowls, uterus, intestine, stomach, spleen, pancreas and bladder, heart, tongue, lungs, liver, kidneys and inspection of the carcass. These procedures are

inspections are carried out visually, by palpation and by cutting where necessary, using the appropriate equipment and facilities. The inspection must avoid contaminating carcasses that are visually in good sanitary condition with those that are contaminated or have lesions suggestive of disease (BRASIL, 1995). The jowls should be opened and the masseter and pterygoid muscles cut so that the lymph nodes in the area remain intact, in order to prevent pathogenic bacteria such as *Yersinia enterocolitica* and *Salmonella* sp., present in lymph nodes and tonsils, are not spread during this procedure (DE BUSSER et al., 2011; FREDRIKSSON-AHOMAA et al., 2009; FRONDREVEZ et al., 2010; VIEIRA-PINTO et al., 2012).

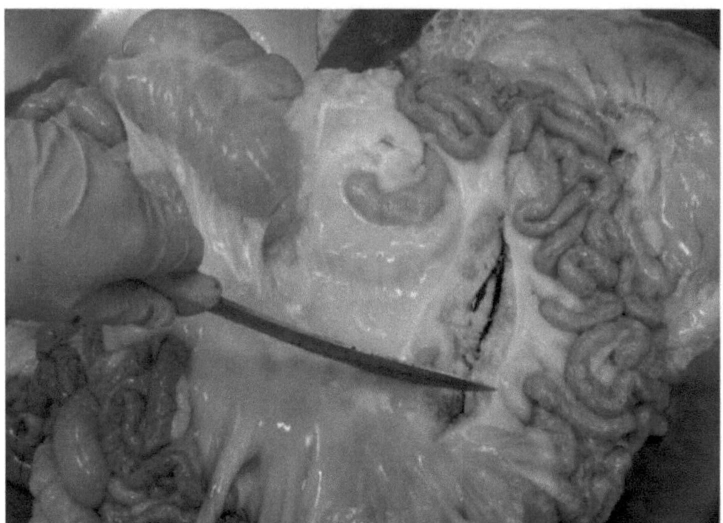

Figure 5 - Post-mortem inspection of pig intestines with mesenteric lymph nodes cut out
Source: Rost (2008)

Once the *post-mortem* inspection procedures have been completed, the carcasses go to the toilet, which includes removing the bleeding wound, the spinal cord and the lymph nodes, followed by washing the carcasses with water at room temperature. Washing is carried out with the carcass passing through the shower, located at the end of the clean slaughter area. The shower lasts around 3 to 4 seconds. This shower must be built in the form of a metal box, made of stainless steel, with a width of 1.60 metres and a minimum height of 4 metres,

length according to the hourly slaughter speed, with 3.60 metres for slaughter speeds of up to 300 pigs per hour. The water must be in the form of jets in sufficient volume and with a pressure of 3atm (three atmospheres), coming from tubular hydraulic installations located in the upper, middle and lower parts of the box. The waste water must be drained directly into the sewage system, using its own pipework (BRASIL, 1995).

After the bath (Figure 6), the carcasses go on to be cooled, which takes place in cold rooms with a controlled temperature of between -1°C and 1°C and an air speed of 2 to 3 metres per second. This stage ends the operations of the clean slaughter zone and the carcasses only leave the chambers when the internal temperature measured inside the leg reaches at least 7°C (BRASIL, 1995). Refrigeration prevents the multiplication of mesophilic pathogens and reduces the multiplication of psychotrophic pathogens and spoilage microorganisms. The low air temperature inside the chambers, combined with the air speed and relative humidity, causes rapid water loss from the surface of the carcasses, making them drier. The cooling of the surface and the low water activity (aw) slows down the multiplication of some pathogens and is lethal for others (INTERNATIONAL COMMISSION ON MICROBIOLOGICAL SPECIFICATIONS

FOR FOODS, 2005). This can be explained in terms of thermodynamics, by the decrease in entropy. The result is an increase in the concentration of compatible solutes, in response to the *stress of* low temperature and osmosis, hampering the homeostasis of the cell cytoplasm, the conservation of energy and, consequently, bacterial survival and multiplication (MCMEEKIN et al., 2010).

Several authors emphasise the importance of the pig's health *status* in bacterial transmission during health inspection procedures, due to the cuts made to the lymph nodes and concomitant manipulation of the organs, leading to cross-contamination of the carcasses at slaughter (BESSA; COSTA; CARDOSO, 2004; HUMPHREY; JORGENSEN, 2006; OLIVEIRA et al., 2010; SILVA et al., 2009).

Figure 6 - Shower for washing pig carcasses located at the end of the clean slaughter area
Source: Cattani (2010)

The European Community's Scientific Committee on Veterinary Measures relating to Public Health has issued an opinion on the revision of meat inspection procedures, after being asked to review the *ante* and *post-mortem* inspection steps for pigs provided for in Directive 64/433/EEC and Directive 91/497/EEC. The committee concluded that the current inspection procedure has major limitations for preventing zoonotic infections in humans. They found no risk to human health through transmission from lesions found on slaughtered pigs during the observation and palpation procedures at *post-mortem* inspection. There is, however, the risk of cross-contamination by pathogens present through the handling and incision of carcasses and organs during *post-mortem* inspection (EUROPEAN COMMISSION, 2000).

2.3 MICROBIOTA   OF PIGS

The possibility of microbial contamination of the surface of pig carcasses in a slaughterhouse is wide-ranging, both by deteriorating and pathogenic bacteria (LIMA et al., 2004). It is therefore necessary to know the microbiota present in pigs.

Potential *pathogens* such as *Clostridium perfringens, Listeria monocytogenes and Yersinia enterocolitica* are found throughout the pig's digestive tract, which includes the nasopharynx and intestines,

Salmonella spp., *Escherichia coli* O157:H7 and *Campylobacter* spp. The mesenteric lymph nodes can contain *Salmonella* spp., *Mycobacterium* spp., *Rhodococcus equi and Yersinia enterocolitica*. The skin is a source not only of *Staphylococcus aureus* and *S. hyicus,* bacilli and other mesophiles, but also of MICRO-organisms of faecal origin as a result of contamination acquired on the farm, during transport or in slaughterhouse pens, such as *Salmonella* spp. (CORNICK; HELGERSON, 2004; LARA et al, 2011; MATARAGAS; SKANDAMIS; DROSINOS, 2008; NESBAKKEN et al., 2003; VELOSO, 2000).

In an investigation carried out by Lara et al. (2011) at slaughter, in a sample of 258 submaxillary and mesenteric porcine lymph nodes (129 with lesions and 129 without apparent lymphadenitis lesions), *Mycobacterium* spp. were found, *Rhodococcus equi, Staphylococcus* sp., *Streptococcus* sp., *Enterobacter cloacae, Proteus mirabilis, Escherichia coli* and other enterobacteria were found. These microorganisms were isolated in 105 (81.4%) of the 129 lymph nodes with lesions and 34 (26.3%) of those without. Similarly, Nesbakken et al. (2003) investigated the occurrence of *Yersinia enterocolitica* and *Campylobacter* spp. in the lymphoid tissues and intestinal tract of pigs, as well as the risk of contamination during *post-mortem* inspection and concluded that the incision of the submaxillary lymph nodes represents a risk for cross-contamination by *Yersinia*, as does the incision of the mesenteric lymph nodes. *Campylobacter* spp. was found in 100 per cent of the intestinal tract of the slaughtered animals and in 66.7 per cent of the tonsils.

In Brazil, Bessa, Costa and Cardoso (2004), studying the prevalence of *Salmonella* spp. in 300 pigs slaughtered in slaughterhouses in Rio Grande do Sul, found *Salmonella* spp. in 167, of which 53 had *Salmonella* spp. only in the mesenteric lymph nodes, 55 only in the faecal content and 59 in both materials collected. In the state of Mato Grosso do Sul, Silva et al. (2009) collected 300 samples of mesenteric lymph nodes and tonsils from slaughtered pigs and found 50 samples positive for *Salmonella* spp. and 14 different serovars.

These studies demonstrate the diversity of pathogens found during slaughter procedures and the importance of managing the risks associated with meat safety by developing pathogen control strategies at all stages of processing (SOFOS; GEORNARAS, 2010).

2.4 INDICATOR MICROORGANISMS

Indicator microorganisms are used to assess the effectiveness of procedures that influence the microbiological quality of food.

2.4.1 Aerobic mesophilic microorganisms

Mesophilic microorganisms account for the vast majority of those important in food, including most of the pathogens in food. They have an optimum multiplication temperature of between 25°C and 40°C, a minimum of between 5°C and 25°C and a maximum of between 40°C and 50°C. The count is commonly used to indicate the sanitary quality of food. Even if pathogens are absent and no changes in sensory characteristics have occurred, a high number of microorganisms indicates that the food is unhealthy. An exception should be made for fermented foods (FRANCO; LANDGRAF, 2005).

To count total mesophiles, the depth plating method on standard counting agar is used. The method is based on counting the number of viable cells, i.e. the ability of a cell to divide and multiply in a suitable culture medium. The group of total mesophiles includes aerobic, anaerobic and facultative aerobic mesophilic micro-organisms (BRASIL, 2003).

2.4.2 Micro-organisms from the *Enterobacteriaceae* family

Members of the *Enterobacteriaceae* family are not confined solely to the intestinal tract and can be isolated from a variety of non-intestinal sources. These members are good indicators of the environmental hygiene process because they are readily inactivated by sanitisers and capable of colonising a variety of niches in the processing plant when sanitisation is inadequate (KORNACKI; JOHNSON, 2001).

In Brazil, enterobacteria and *Escherichia coli* are used as indicator microorganisms to determine the hygiene of the pig slaughter process. They are collected by *swabbing* the surface of the carcass in exporting establishments (BRASIL, 2007).

This family includes the genera *Citrobacter, Enterobacter, Erwinia, Escherichia, Hafnia, Klebsiella, Proteus, Providencia, Salmonella, Serratia, Shigella* and *Yersinia* (KORNACKI; JOHNSON, 2001).

In Europe, enterobacteria have been used for years as a quality indicator, while in the US, the coliform group has traditionally been used. The coliform group is defined on the basis of their biochemical reactions and not on the basis of genetic relationships; they refer to bacteria in the form of bacilli that ferment lactose, forming acid and gas within 48 hours at 35°C. The so-called faecal coliforms or thermo-tolerant coliforms are defined as those that ferment lactose, forming acid and gas within 48 hours, at 44.5° and 45.5°C, usually in EC broth (KORNACKI; JOHNSON, 2001).

Enterobacteria are Gram-negative, facultative anaerobic bacteria that ferment glucose into acid, are oxidase negative, usually catalase positive, normally reduce nitrate and may or may not be mobile (KORNACKI; JOHNSON, 2001). Counting is based on inoculating the desired dilutions of the sample onto crystal violet red neutral bile glucose agar (VRBG), the composition of which shows the ability of the microorganisms to ferment glucose with the production of acid, a reaction indicated by the indicator turning red and the precipitation of bile salts around the colonies. Selectivity is exerted by the presence of crystal violet and bile in the medium. After growth, the colonies are confirmed by the oxidase test. In this test, the appearance of a blue colour (N'N'N'N'-tctramctyl-parafcnilcno- diamine) or an intense red colour (para-amino-dimethylaniline oxalate) is indicative of a positive reaction. All enterobacteria show a negative oxidase reaction (BRASIL, 2003).

2.4.3 *Escherichia coli*

Escherichia coli belongs to the *Enterobacteriaceae* family *and is* a normal component of the intestinal microbiota of warm-blooded animals, including humans (ACHA; SZYFRES, 2003). Commensal strains of *E. coli* rarely cause disease, except in immunologically compromised animals or when the normal gastrointestinal barrier is breached (MENG et al., 2007).

Escherichia coli is a Gram-negative, facultative anaerobic bacterium. In biochemical reactions, it has the following profile: produces acid and gas from glucose metabolism, oxidase negative, catalase positive, methyl red positive and Voges-Proskauer reaction negative; usually citrate negative. Does not produce hydrogen sulphide, does not hydrolyse urea or lipase. Reduces nitrate. Produces indole from

of tryptophan metabolism and produces the enzyme P-glycorunidase (BERGEY et al., 1994).

The activity of the P-D-glycorunidase enzyme is specific to *E. coli* and some strains of *Shigella*. The 4-methylumbelliferyl -P-D- glycorunid (MUG), when incorporated into the *E. coli* culture broth, will be cleaved by the enzyme present in these strains into a compound called 4-methylumbelliferone that blooms when exposed to long wavelengths. It is important to remember that enterohaemorrhagic strains of *E. coli,* including O157:H7, do not have this enzyme (BLOOD; CURTIS, 1995). *Escherichia coli* is considered the best indicator of faecal contamination for raw foods (KORNACKI; JOHNSON, 2001).

The identification of *E. coli* is based on inoculating the desired dilutions of the samples on crystal violet red neutral bile agar (VRBA) and then counting the suspected colonies. Crystal violet red neutral bile agar contains bile salts and crystal violet, which are responsible for inhibiting Gram-positive micro-organisms, and neutral red, a pH indicator that reveals the fermentation of lactose by the micro-organisms present. The addition of an overlay after the inoculum has been homogenised with the agar aims to prevent the growth and spreading of colonies on the surface. Suspect colonies are inoculated into EC Broth with Durhan tubes. The presence of gas in the Durhan tubes indicates fermentation of the lactose present in the medium. *Escherichia coli is* tested by transferring a portion of the positive tubes of EC broth to an Eosin Methylene Blue (EAM) plate, which is evidenced by the presence of purple colonies with a dark centre and a characteristic greenish metallic sheen, after incubation at 36 °C for 24 hours. Confirmation of *Escherichia coli* is based on biochemical tests (BRASIL, 2003).

2.5 PATHOGENIC MICRO-ORGANISMS

There are countless pathogenic microorganisms that can cause harm to public health. This study will only look at two enterobacteria classified as pathogenic.

2.5.1 *Salmonella* spp.

Salmonella spp. is a member of the *Enterobacteriaceae* family responsible for infections in pigs and humans, which can influence meat production and public health.

2.5.1.1 Classification

The genus *Salmonella* consists of two species: *S. enterica* and *S. bongori.* *Salmonella enterica* has six subspecies: Salmonella enterica subsp. Enterica, also classified as I; Salmonella *enterica* subsp. Salamae, or II; Salmonella *enterica* subsp. Arizonae, or Illa; *Salmonella enterica* subsp. Diarizonae, or Illb; *Salmonella enterica* subsp. Houtenae, or IV; *Salmonella enterica* subsp. Indica, or VI and numerous serotypes based on the serological characteristics of each subspecies (D'AOUST; MAURER; BAILEY, 2007). Only twenty-one serovars belong to the *Salmonella bongori* species, also known as subspecies V, and a new serovar was reported in 2005 (HERRERA-LEÓN et al., 2005).

According to the Kauffmann-White classification, *Salmonella* is divided into serotypes, distinguished by their antigenic structure, made up of O somatic, H flagellar and Vi capsular antigens (D'ÁOUST; MAURER; BAILEY, 2007). Pigs are hosts to numerous *Salmonella* serotypes and are considered to be the main reservoir of *Salmonella cholerasuis,* responsible for necrotic enteritis, which is common in herds living in poor hygiene conditions. The serotypes that attack pigs include *S.* Enteretidis, *S.* Typhimurium and *S.* Dublin. These serotypes are usually isolated from the intestine and mesenteric lymph nodes. Due to the frequency of infection of pigs with different *Salmonella* serotypes, products derived from these animals have commonly been a source of human infection (ACHA; SZYFRES, 2003). In general, pigs infected with these *Salmonella* serotypes are healthy carriers, considered asymptomatic carriers of public health importance (FEDORKA-CRAY; GRAY; WRAY, 2000).

2.5.1.2 Ecology

Salmonella infections are recognised as zoonoses and its habitat is the intestinal tract of mammals and birds (warm-blooded animals), but it is also found in some cold-blooded animals, such as reptiles (ACHA; SZYFRES, 2003). More than 60 per cent of all strains identified and 99 per cent of the serotypes responsible for diseases in warm-blooded animals are members of subspecies I. The other *Salmonella subspecies,* in particular subspecies Illa and *Salmonella bongori,* are associated with diseases in cold-blooded animals (HERRERA-LEÓN et al., 2005).

Salmonella spp. can remain in transport lorries, pens, floors and equipment at the slaughterhouse when excreted from batches of seropositive pigs, even after cleaning and disinfection procedures (SWANENBURG et al., 2001). It can persist in the pig for a long time, especially in the digestive tract and the corresponding lymphatic tissue, mainly in the tonsils, mesenteric and mandibular lymph nodes. The presence of *Salmonella* spp. in the mandibular lymph nodes (which are normal in appearance) constitutes a risk because this tissue can remain in the head - with the carcass - after slaughter and also because they are incised during health inspection, allowing cross-contamination via the inspector's knife (VIEIRA-PINTO et al., 2010). The process of evisceration, toileting and *post-mortem* inspection together result in a 5% to 10% increase in *Salmonella* spp. on pig carcasses (BERENDS et al., 1997).

Salmonella spp. has the ability to persist in hostile environments and for prolonged periods of time (D'ÁOUST; MAURER; BAILEY, 2007), becoming the resident microbiota in slaughterhouses (SOFOS; GEORNARAS, 2010). Pork and its products are contaminated with *Salmonella* during slaughter and processing, through infected and asymptomatic pigs or subclinical infections that reach the slaughterhouse, which is a high risk factor (GUENTER et al., 2010).

2.5.1.3 Prevalence

In Denmark, from 13468 finishing pigs (ready for slaughter), 30 different serotypes of *Salmonella enterica* were isolated from 832 pigs (6.2 per cent), with *Salmonella* Typhimurium being the predominant serotype in 536 (64.4 per cent) of the isolates (BAGGESEN et al., 1996).

Studies to determine the prevalence of *Salmonella* spp. in pigs slaughtered in southern Brazil were carried out by Bessa, Costa and Cardoso (2004), who found prevalence rates ranging from 55.6%, with the serotype most commonly found being Typhimurium, in 24.36% of the samples, followed by Agona (19.9%), Derby (13.2%) and Bredeney (12%). The Typhimurium and Derby serotypes were the ones predominantly found in a study carried out by Seixas, Tochetto and Ferraz (2009) in Santa Catarina. Schwarz et al. (2009) found a prevalence of 73.8% to 83.2% in the blood and lymph nodes of the pigs sampled, with isolation of 62.5% to 85.0%.

In an experiment conducted in the state of São Paulo by Jakabi et al. (2004), *Salmonella enteritidis* was the serotype most commonly found (8/23) and 9.1% (23/253) of the meat samples were positive for *Salmonella* spp. This percentage was distributed in 19.6% of raw pork sausages, 8.7% of raw pork, 7.3% of raw beef and 4.6% of poultry, which demonstrates the importance of controlling the pathogen before the carcass reaches refrigeration. In pork products, such as fresh sausage, Spricigo et al. (2008) found 27% of *Salmonella* spp. in 200 samples collected in the state of Santa Catarina, with the Derby serotype being the most commonly found. Similarly, Almeida et al. (2000), analysing 169 food samples, found 59.7% of *Salmonella* in meat and meat products in the state of São Paulo, with the Enteritidis serotype being the most commonly found. Murmann, Santos and Cardoso (2009), analysing 336 samples of fresh sausage in the state of Rio Grande do Sul, detected 24.4% with *Salmonella enterica*. In the state of Santa Catarina, 2492 samples of pork and pork products were analysed between 2007 and 2009. *Salmonella* spp. was found in 6.06% of them, of which 7.11% were *fresh* pork cuts and trimmings (CATTANI, 2011).

In the European Union, 131,468 cases of food-borne human salmonellosis were reported in 2008 (SCIENTIFIC...., 2010). In the USA, 5,267 cases were confirmed by the CDC in 2008 (UNITED STATES OF AMERICA, 2008). In Brazil, 1275 cases of *Salmonella* spp were reported between 1999 and 2008, with the southern region of Brazil reporting the most cases (BRASIL, 2010).

2.5.1.4 Cultural characteristics and identification

Salmonella spp. are Gram-negative, rod-shaped, facultative anaerobic bacteria that multiply at an optimum temperature of 37 °C and belong to the *Enterobacteriaceae* family. As biochemical characteristics, they catabolise D-glucose and other carbohydrates with the production of acid and, usually, gas. They are oxidase negative and catalase positive, indole and Voges Proskauer negative, methyl red and Simons citrate positive. They use citrate as a carbon source; they generally produce hydrogen sulphide with decarboxylation of lysine and omitidine and do not hydrolyse urea (BERGEY et al., 1994).

After pre-enrichment, the identification of *Salmonella* spp. is carried out with selective enrichment, with the aim of inhibiting the multiplication of the accompanying microbiota and preferentially promoting an increase in the number of *Salmonella* cells. Two enrichment media are used, since *Salmonella* resistance to selective agents varies depending on the strain: tetrationate broth (TTB) and Rappaport Vassiliadis broth (RP).In Rappaport Vassiliadis broth, temperature, combined with the presence of malachite green and magnesium chloride, act as selective agents for the accompanying flora, while the presence of peptone stimulates *Salmonella* multiplication.

In tetrationate broth, micro-organisms that have the enzyme tetrationate reductase, such as *Salmonella* and *Proteus,* multiply, while at the same time inhibiting the multiplication of other micro-organisms of faecal origin that don't have this enzyme.

Brilliant green and iodine inhibit the multiplication of Gram-positive microorganisms. Plating is then carried out on selective differential media in order to obtain isolated *Salmonella* colonies with typical characteristics that are distinct from those of competitors, for subsequent serological and biochemical confirmation.

Hektoen enteric agar (HEA) is used as a differential selective medium because, due to the action of the indicators bromothymol blue and acid fuchsin, it shows a significant colour difference between lactose-positive and lactose-negative colonies. Thiosulphate, by combining with iron ions, acts as an indicator in the production of H2S, observed by the blackening in the centre of the colonies. The presence of bile salts inhibits a large part of the accompanying flora.

Plating is also carried out on xylose lysine deoxycholate agar (XLD), which inhibits the growth of Gram-positive microorganisms due to the deoxycholate present in the medium. Xylose is fermented by practically all enteric microorganisms, except *Shigella*. The decarboxylation of lysine by the *Salmonella* group turns *Salmonella* colonies red due to the alkalisation of the medium, allowing differentiation from other micro-organisms that produce yellow colonies by fermenting the lactose and sucrose present. Sodium thiosulphate and ammoniacal ferric citrate are also found, which allow the production of $H_2 S$ by *Salmonella,* resulting in colonies with a black centre. Colonies with typical growth are screened with a lysine decarboxylation test (LIA), fermentation of lactose and/or sucrose, production of $H_2 S$ on Iron Lysine Agar and Iron Triple Sugar Agar (TSI) and a urea test; if negative for *Salmonella,* the subsequent steps can be eliminated.

After screening, they are submitted to the polyvalent somatic serological test (slide agglutination technique), which is based on the antigen-antibody reaction, with consequent agglutination of the antigen against the antibody, to identify *Salmonella* O, K and H antigens. For biochemical confirmation, the dulcitol, lactose, sucrose, indole, malonate, citrate, hydrocyanic acid (KCN), methyl red and Voges-Proskauer fermentation tests are carried out (BRASIL, 2003).

2.5.2 E. coli O157:H7

E. coli is a member of the *Enterobacteriaceae* family, whose pathogenic strains have become important for public health since the 80s.

2.5.2.1 Classification

E. coli, a member of the *Enterobacteriaceae* family, is one of the most important enteric pathogens, along with *Salmonella, Shigella and Yersinia.* Although many *E. coli* do not cause gastrointestinal illness, some groups can cause life-threatening diarrhoea (MENG; ZHAO; DOYLE, 2001). *E. coli* is classified into different serotypes according to the classification originally developed by Kauffmann, which is based on the identification of the somatic antigen O (polysaccharide and thermostable), which differentiates *E. coli* into more than one hundred and seventy different types. *coli* into more than one hundred and seventy serogroups; the flagellar antigen H, derived from the flagella of mobile strains, which is thermolabile and is the flagellar protein that carries the antigenic factors determining group H, with more than fifty-three groups; the antigen K, which is a capsular polysaccharide and F, fimbrial antigen, for strains that have this structure. Strains without flagella are identified as NM *(nom motile)* (ACHA; SZYFRES, 2003; MENG et al., 2007).

Pathogenic strains of *Escherichia coli* cause diarrhoea and are classified into specific groups. The classification is based on virulence properties, pathogenicity mechanisms, clinical syndromes and distinct O:H serotypes. This category includes enteropathogenic *E. coli* (EPEC), enterotoxigenic E. coli (ETEC), enteroinvasive E. coli (EIEC), diffuse-adherent E. coli (DAEC) and enterohaemorrhagic E. coli (EHEC), which is the most important group, based on the frequency and severity of the disease (MENG et al., 2007).

The enterohaemorrhagic *E. coli* (EHEC) group includes *E. coli* O157:H7 as the most prevalent serotype, as well as other non-0157 serotypes that cause haemorrhagic colitis. All EHEC produce cytotoxicity factors in Vero cells, called verotoxin-producing (VT) or *shiga-like* toxins (SLTs), which are toxic proteins, biologically, structurally and antigenically related to the cytotoxin called shiga-toxin (ShT), produced by *Shigella dysenteriae* type 1 (AGBODAZE, 1999, MENG et al., 2007).

E. coli of other serotypes have also been shown to produce verotoxins (VT) or *shiga-like* toxins (SLTs) and have therefore been called Shiga toxin-producing *E. coli* (STEC) or Verotoxigenic *E. coli* (VTEC), associated with haemorrhagic colitis (HC) and Haemolytic Uremic Syndrome (HUS) in humans. Other non-0157 strains that produce haemorrhagic colitis in humans have been identified in several countries and include the serotypes: 026, 0111, 0103, 0121, 045, 0145, 091, 0118, 0119 and 0128 (ACHA; SZYFRES, 2003; LEE et al., 2009; MENG et al., 2007; OJO et al., 2010).

E. coli O157:H7 strains can carry various virulence factors, such as the *slxl* and *stx2* genes, which produce Shiga toxins, and the *eae* gene, which causes an important adherence effect on cells by encoding intimin, called *"attaching and effacing"*, as well as the *hlyk* gene, which has haemolytic activity. It also has a plasmid called pO157 of approximately 60 MDa (megadaltons), which is implicated in the pathogenesis of the infection (ATEBA; MBEWE, 2011; HEUVELINK et al., 1999; MENG; ZHAO; DOYLE, 1998; WANI et al., 2009; WONG; MACDIARMID; COOK, 2009).

2.5.2.2 Ecology

In oedema disease in piglets and haemorrhagic colitis in lambs, *Shiga-like* toxins have been considered as a causative agent of these diseases (AGBODAZE, 1999). When pathogenic strains of *E. coli* adhere and multiply in the small intestine of susceptible piglets, they produce SLTs, which are absorbed. They then enter the bloodstream, where they cause systemic vascular injury, damaging the endothelium and tunica media of small arteries and arterioles throughout the body, which leads to clinical signs of motor incoordination and ataxia. Oedema can be seen on the face and eyelids while the animal is still alive and at necropsy, involving internal organs such as the intestine. Diarrhoea is not a common symptom (JONES; HUNT; KING, 2000; SOBESTIANSKY et al, 1999).

Enteric colibacillosis complicated with shock also occurs in young pigs, before and after weaning. The *E. coli* associated with this disease can be ETEC and commonly belong to serogroups 0149, 0157 and S8 - which are F4 positive, produce thermostable enterotoxins (ST-I and ST-II) and rarely thermolabile toxins (LT-I), but occasionally produce *Shiga-like* toxin (SLT-lle) or only produce SLT-lle associated with oedema disease (FAIRBROTHER, 1999).

2.5.2.3 Prevalence

In Brazil, a study carried out in 2001 on the diarrhoeal faeces of 34 piglets with symptoms of oedema disease found 144 strains of *E. coli*, 99 of which had the *stx2e* gene and 41 of which belonged to serogroup 0139 (SILVA, A. et al, 2001).

In Denmark, the majority of ETEC and VTEC strains involved in oedema disease and post-weaning diarrhoea in pigs belong to a limited number of serotypes, with 0149 being the most prevalent, found in 49.9% of the strains analysed. The 0157 was found in 0.9% of 563 *E. coli* strains and the majority of all isolates had haemolytic activity (FRYDENDAHL, 2002).

The prevalence of *E. coli* O157:H7 was higher in pig faeces than in cattle and human faeces - a result reported by Ateba and Bezuidenhout in 2008 in southern Africa - which suggests that these animals could be a risk to human health by carrying this strain.

In Japan, 1.4% of the pigs sampled from 35% of the country's farms had STEC O157:H7, carried the *stxi, stX2 , eaeA* genes and harboured the pO157 plasmid (NAKAZAWA; AKIBA; SAMESHIMA, 1999). In a study to determine the variants of the *stx* gene$_2$ in STEC strains isolated from human patients in Japan, the variants were grouped according to the nucleotide sequence of the Stx2 family *and* a significant relationship was found between Stx2 and HUS (NAKAO et al, 2002).

In 2007, an outbreak of *E. coli* 0157 was reported, associated with the consumption of pork salami, where the *slxj, stX2 and eae* genes were isolated from patients hospitalised with haemorrhagic colitis and from the salami consumed (CONEDERA et al, 2007). In 2008, the CDC reported

532 cases of *E. coli* O157:H7 in the USA (UNITED STATES OF AMERICA, 2008). Brazil does not have systematised data on the number of cases of HUS or CH caused by pathogenic strains of *E. coli* O157:H7.

2.5.2.4 Cultural characteristics and identification

The O157:H7 serotype is the only one of the majority of *E. coli* strains that does not ferment sorbitol within 24 hours and does not express the 0-glucoronidase found in many *E. coli* strains. Thus, Sorbitol MacConkey Agar (SMAC) has been chosen for the isolation of O157:H7 in food, with the addition of cefixime-telurite, in order to inhibit other strains of *E. coli* and other bacteria that do not ferment sorbitol (MENG; ZHAO; DOYLE, 2001).

E. coli O157:H7 strains are unable to multiply at temperatures above 44 to 45.5°C; they are therefore inhibited and are not detected in faecal coliform analyses using the most probable number method, which uses lactose fermentation at 44.5°C as a confirmatory characteristic, nor in direct *E. coli* analyses using substrates for the 0-glucoronidase enzyme (KORNACKI; JOHNSON, 2001; SILVA et al., 2003).

2.6   CARCASS DECONTAMINATION METHODS

Improving the microbiological quality of food by improving the technological process is not enough to guarantee that food-borne illnesses do not reach the consumer, because the technological process does not always guarantee the absence of pathogens. Food can easily be re-contaminated. Efforts must be made to strictly comply with hygiene measures, following Good Hygiene Practices (GHP) and Good Manufacturing Practices (GMP), rigorously implemented in the Hazard Analysis and Critical Control Points (HACCP), throughout the food chain (PANISELLO et al., 2000). Carcass decontamination is a relevant option in cases where the prevalence of pathogens in carcasses must be reduced to levels that cannot be achieved by implementing hygiene practices at slaughter and interventions at primary production level (LAWSON et al., 2009).

Bacterial levels can be reduced in the slaughter process with prior care, through modified diets, competitive exclusion, using bacteria beneficial to the animal, the use of treatments in drinking water, oral administration of chlorates and the use of vaccines. After slaughter, it is also possible to use decontamination techniques such as chemical hair removal, washing with hot water at temperatures >74 °C, pasteurisation with steam, vacuuming with steam, chemical washing with organic acid solutions and the use of lactoferrin as an antimicrobial in food (HUFFMAN, 2002; KOOHMARAIE et al., 2007).

The outbreaks of *E.coli* O157:H7 in the United States in 1993 ushered in an era of intensive efforts to improve the sanitary quality of meat. Various types of interventions, such as the use of organic acids, hot water, testing and retesting of samples, as well as irradiation, were carried out to ascertain the best and most effective method of reducing pathogenic micro-organisms (KOOHMORAIE et al., 2005).

In early 1996, the US Food and Drug Administration (FDA) approved the use of acidified sodium chloride as a permitted additive to reduce pathogens and extend the shelf life of red meat, poultry and seafood products (UNITED STATES OF AMERICA, 2011). In the European Community, the use of decontamination of fresh pork is not common. For many years, European legislation limited the use of substances for the surface removal of animal products. However, Regulation 853/2004 EC currently provides the legal basis for the use of physical methods, such as hot water or steam, as a decontamination method (CODE...,2004b; HUGAS; TSIGARIDA, 2008).

In Brazil, the use of chemical substances in the process of microbial decontamination of carcasses was carried out until 2004 in slaughterhouses under federal inspection by the Ministry of Agriculture. However, in view of the possibility of exporting pork carcasses to member countries of the European Union, and since this practice is condemned by these countries, in 2004 the Department of Inspection of Products of Animal Origin of the Ministry of Agriculture decided to suspend this practice until further studies on the subject were carried out in Brazil (BRASIL, 2004). Studies have been carried out in various countries to verify the efficiency of sanitising the carcasses of different animal species, with organic, acetic and lactic acids being the most commonly used in the process of decontaminating and washing carcasses, associated or not with the use of hot water and steam (BOSILEVAC et al., 2006; CARPENTER; SMITH; BROADBENT, 2011; DORSA et al., 1995; EPLING; CARPENTER; BLANKENSHIP, 1993; HARDIN et al., 1995; HUFFMAN, 2002; HWANG; BEUCHAT, 1995; NISSEN; MAUGESTEN; LEA, 2001).

2.6.1 Organic acids

Organic acids and their salts are considered weak acids. Acetic, lactic, propionic, benzoic and sorbic acids, which are also considered weak acids, are commonly used in food preservation (DAVIDSON; TAYLOR, 2007), inhibiting multiplication, reducing the number and prevalence of pathogens, as well as the microbial load of carcasses (HUFFMAN, 2002).

The effectiveness of the antimicrobial activity is related to the pH and the undissociated form of the acid. In this way, the undissociated acid permeates through the plasma membrane (lipolytic), penetrating the cell. The acid molecule dissociates after entering the cell, resulting in the release of anions and protons, which acidify the cell cytoplasm and must be expelled to the outside. However, the membrane is impermeable to the protons that must be transported to the outside. This leads the cell to a homeostatic imbalance, creates an electrochemical potential across the membrane, alters the intracellular pH, inhibits essential metabolic reactions and accumulates toxic substances (DAVIDSON; TAYLOR, 2007; FORSYTHE, 2002).

In order to be considered safe for use as antimicrobials, organic acids must meet certain requirements: they must not have a permanent action on the meat; the colour of the fresh meat must not be preserved; the final product must show indicators of normal deterioration (for example, discolouration); there must be no extension of the shelf life compared to products made from untreated meat; the nutritional composition of the meat must not be affected by the treatment - for example: proteins must not be denatured; the sensory characteristics of the product must not be altered in relation to the final product and, finally, there must not be any detectable residues of the organic acid in the meat product made from the treated carcass (UNITED STATES OF AMERICA, 2011).

The effects of microbiological decontamination of bovine carcasses were analysed by Gill and Landers (2003), using 2% lactic acid, washing the carcasses with water between 40°C and 50°C and pasteurisation with steam or hot water at 85°C, the latter being considered the most efficient in achieving maximum bacterial reduction. A commercial solution of 5% chlorine dioxide was used on the surface of pig carcasses by Sardinha et al. (2001), and was found to significantly reduce enterobacteria and the total aerobic bacteria count. Similarly, studies carried out by Vasconcelos et al. (2002) on sheep meat using 1% acetic acid inhibited the growth of mesophiles, moulds and yeasts for a fortnight and coliforms for six weeks.

Studies were carried out by Stopforth et al. (2003) to evaluate the influence of the concentration of organic acids on the survival of *Listeria monocytogenes* and *Escherichia coli* O157:H7 in carcass washing water and on equipment surface models. They showed that the pH of the washing water, around 4, was lethal for *Listeria monocytogenes,* but not for *Escherichia coli* O157:H7.

Organic acids are excellent antimicrobials against bacteria, including *Salmonella.* They offer several advantages as antimicrobials because they are GRAS *(Generally Regarded as Safe)* for use without limitation, they are low cost, easily manipulated and cause minimal sensory changes in products (MANI-LÓPEZ; GARCIA; LÓPEZ-MALO, 2012).

Recent studies suggest that there is an influence of the low pH attributed to the organic acids produced by probiotic bacteria in inhibiting the expression of the *stx2A* gene *of E. coli* O157:H7 (CAREY et al., 2008). At the same time, there is concern that decontamination treatments may lead to increased tolerance after adaptation to environmental stress and microorganisms may become highly resistant to other stress situations, resulting in increased survival of pathogens in the cams (STOPFORTH et al., 2003; SMIGIC et al., 2009). To assess the impact of decontamination on public health, some aspects need to be considered, such as the prevalence of pathogens in carcasses, the distribution of the number of pathogens in each carcass, the detection limit for microbiological methods - which should be a fixed number or the probable distribution - the infective dose and type of the pathogen and the conditions along the chain, which determine the multiplication, survival or death of the bacteria after decontamination (HUGAS; TSIGARIDA, 2008).

2.6. 2Lactic acid

Lactic acid (2-hydroxypropanoic acid) (Figure 7) is a monocarboxylic acid with a pKa of 3.79 and is a normal metabolite in humans and animals. It is oxidised by pyruvate in the cytoplasm of the cell to carbon dioxide and water in the tricarboxylic acid cycle inside the mitochondria. It is produced naturally during the fermentation of food by lactic acid bacteria. Lactic acid and its salts are powerful disintegrators of the outer membrane composed of lipopolysaccharides (LPS), altering the permeability of the barrier (DAVIDSON; TAYLOR, 2007).

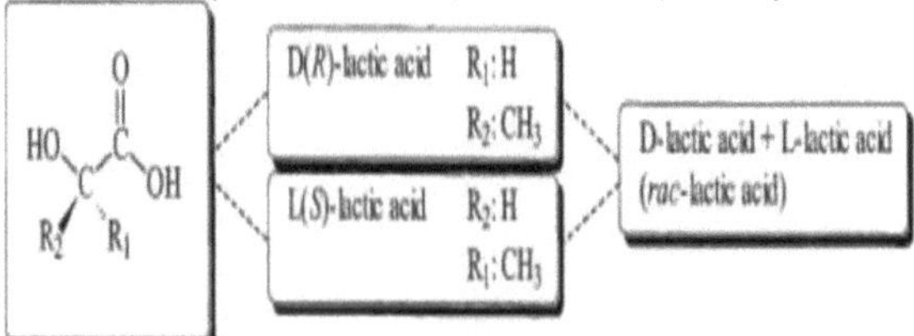

Figure 7 - Lactic acid stereoforms
Source: Sodergard and Stold (2002)

In the USA, the use of this substance is regulated by the Food Safety and Inspection Service (FSIS) of the United States Department of Agriculture (USDA), through FSIS Directive 7120.1, revision 11 of 4/6/2012, which establishes limits of use and applications. It can be used in an aqueous solution for spraying, at a concentration of up to 5%, on carcasses before or after refrigeration and on various organs and meats. This directive also establishes certain requirements for organic acids to be approved and used in industry (UNITED STATES OF AMERICA, 2012).

Using low dilutions of lactic acid, between 1% and 2%, on the slaughterhouse production line can help in the process of decontaminating carcasses, without affecting the quality attributes of the pork carcass (JAYAWARDANA et al., 2009; VAN NETTEN; MOSSEL; HUIS,1997).

The outer surface of the pork carcass is covered by skin containing stromal protein, mainly collagen, and fat. These components have no apparent colour and are mostly white. The colour of the skin on the surface of the pig carcass at the end of the slaughter line, or

At the time of decontamination, the colour is white-pink-yellow. A slight reddening of the skin may be due to heme pigments from blood residues. Changes in the surface colour of carcasses treated with lactic acid and steam can occur due to the denaturation of the surface layers of the pig's skin, followed by the oxidation of heme pigments and changes in the hydration of proteins due to the reduction in pH after lactic acid treatment (PIPEK et al, 2005).

Lactic acid at 2% and 5%, applied for 60 seconds to the surface of the carcass, eliminated *Salmonella typhimurium* from pig carcasses inoculated with 1 log i_0 CFU/cm, but did not eliminate those inoculated with 2 log$_w$ CFU/cm^2 (VAN NETTEN MOSSEL; HUIS, 1995). The use of 2% lactic acid, associated with water at a temperature of 90°C to 95°C and 4 to 6 bar pressure, was effective in reducing mesophiles and slowing down the growth of psychotrophs (PIPEK et al, 2006).

Intracellular pH has been studied as an indicator of the physiological state of bacterial cells subjected to lactic acid and its interference in cell viability after decontamination (SMIGIC et al, 2009). The effectiveness of washing meat surfaces to decontaminate and inhibit residual multiplication, using 2% levulinic acid, acetic acid and lactic acid, was analysed in a study in which it was found that there was no advantage in reducing pathogens when washing was carried out with levulinic acid, compared to lactic acid and acetic acid (CARPENTER; SMITH; BROADBENT, 2011).

2.6.3 Steam and hot water

The pasteurisation process consists of exposing the carcass or meat products to water vapour at a temperature of between 82°C and 97°C, inside a cabin with atmospheric pressure for 6 to 12 seconds. This treatment usually includes renewable water, pasteurisation with steam and immediate cooling. In some steam equipment, the pressure can be combined to improve treatment efficiency. The industrial process was developed in the USA and the pasteurisation treatment of carcasses was approved by the FDA in 1995 for all carcasses and parts of carcasses that were processed. This technology has the advantage of providing a continuous and inexpensive solution for decontaminating small or large pieces of meat in a short space of time (AYMERICH; PICOUET; MONFORT, 2008).

The effectiveness of the methods used to reduce the number of bacteria on the surface of carcasses is influenced by water pressure, temperature, chemical substances present and their concentration, exposure time, application methods, application chamber model and application stage (BELK, 2002; SOFOS; BELK; SMITH, 1998).

The hot water intervention process has the advantage of achieving a consistent reduction in bacterial contamination and requires less handling of the carcass. Hot water used at a temperature ranging from 75°C to 85°C, for a period of time of 9 to 12 seconds of application, under pressure by spraying on the carcass, is efficient for the decontamination process (BOLTON; DOHERTY; SHERIDAN, 2001).

In an experiment conducted with one hundred and forty-four bovine carcasses, hot water was used at a temperature between 74°C and 87.8°C, concluding that this measure can be an effective intervention to reduce bacteria on the surface of carcasses, as there was a reduction of approximately 1.7 log when compared to the control. This data shows that washing with hot water can be a good strategy and effective intervention to reduce bacteria on the surfaces of bovine carcasses, especially as it produces a low number of bacterial populations between carcasses (REAGAN et al., 1996; KOOHMARAIE et al., 2005). Similarly, Belk (2002) concluded in his work that the use of combined treatments such as: steam, washing with organic acids in pre-evisceration, pasteurisation with hot water, a second application of organic acids and final washing of the carcass, substantially reduced the bacterial count and the incidence of *Salmonella* sp., with a reduction from 14.7% to 1.9%.

Treatment with hot water spray during the first few hours after slaughter significantly reduces the total count of aerobic bacteria, without affecting pork quality (FERREIRA et al., 2004). In pork carcasses, decontamination with hot water resulted in a greater reduction in the number of pathogens (GILL et al., 1995). Carcasses from a slaughter of three hundred and seventy-five pigs per hour, treated for 15 seconds with water at 80°C in a booth, were evaluated and there was a reduction of more than 2 logs in the number of *E. coli*, while refrigeration only reduced the number by 0.7 logs (JENSEN; CHRISTENSEN, 2004).

2.7 MICROBIOLOGICAL DETECTION METHODS

The viable cell count, both in food and on air contact surfaces in a food industry, is one of the most important parameters when it comes to determining food quality (SORRIBES, 2008).

Microbiological detection methods are often categorised into two groups: conventional and rapid. However, the terms are misapplied, as some rapid methods require 24 hours to obtain their results. "Conventional" refers to procedures that are commonly used and usually involve homogenising the food sample, preparing a series of dilutions and inoculating onto specific agar plates for colony formation and subsequent counting. These methods can also be referred to as "traditional". Rapid methods are an alternative to conventional methods and are designed to obtain the final result in less time. This is highly desirable in the food industry, even though the techniques are more expensive (FORSYTHE, 2002).

2.7.1 Conventional or traditional method

Conventional methods, currently used in many laboratories around the world and established as the standard for microbiological food analysis, are characterised by being laborious, using large volumes of culture media and requiring a considerable amount of time to obtain analyses and results (SORRIBES, 2008).

The traditional procedure for counting microorganisms involves several stages: pre-enrichment, selective enrichment, homogenisation, dilution, plating on appropriate selective media, incubation at an appropriate temperature for more than 24 hours and counting specific colonies, followed by biochemical, serological and molecular testing procedures for characterisation (GE; MENG, 2009; OTERO; GARCÍA-LÓPEZ; MORENO, 1998). To detect specific microorganisms, which are often a small proportion of the total microorganisms present in the sample, selective media are used to increase the multiplication of the target microorganism and suppress the rest (GE; MENG, 2009).

In conventional colony counting methods, the number of bacteria in a product is determined by inoculating a diluted suspension of the sample onto the surface of a solid growth medium or by mixing the suspension with liquefied agar in Petri dishes. Counting is completed after incubation at fixed times and temperatures ranging from 7 °C to 55 °C, in an aerobic, micro-aerobic or anaerobic atmosphere, depending on the target micro-organisms. During incubation, each individual cell multiplies into a colony visible to the naked eye. There is, however, the possibility that viable bacterial strains go into dormancy and become unculturable, which can lead to an underestimation of the pathogen or failure to isolate pathogens in contaminated samples. In addition, these methods require two to three days for initial results and seven to ten days for confirmation (JASSON et al., 2010; VELUSAMY et al., 2010).

2.7.2 Rapid or alternative methods

Microbiologists began developing rapid methods in the mid-1960s and accelerated their development from the 1970s onwards due to the need to shorten the time needed to obtain analytical results and improve laboratory productivity (FUNG, 2008).

Alternative methods are defined as those analyses that demonstrate or estimate the same analyte (microorganisms), measured using the corresponding reference method (e.g. AO AC International, FDA/BAM or USDA) and that have the following requirements: rapid analysis and/or responses, easy to operate and/or automate, analytical performance at least as good as those of the reference methods (FELDSINE; ABEYTA; ANDREWS, 2002; LOMBARD; LECLERCQ, 2010).

Rapid or alternative methods require little time to obtain results and allow a large number of samples to be processed per unit of time; they are generally easy to use, accurate (adequate sensitivity and specificity and low detection limits) and cost-effective. It should be noted that, in most cases, the use of rapid methods does not exclude the microorganism enrichment stage (ACHESON et al., 1996; BLACKBURN; MCCARTHY, 2000; SILVA et al., 2001), given the need to obtain pure cultures. Positive results obtained from alternative methods (different from the reference methods) should also be confirmed.

The term "Alternative Method" refers to the combination of products, equipment and complete operating procedures; from sample preparation to the final result (LOMBARD; LECLERCQ, 2010). Alternative methods can be divided into groups, with some systems using one or more types of methodology to achieve the final result: (1) solid culture media, different from those described in the reference methodology; mostly based on chromogenic substrates, making it easy to visually recognise the target bacteria on the plates; (2) immunoenzymatic kits, such as ELISA and (3) molecular methods, such as hybridised probes or PCR, both conventional or *real time* (FUNG, 2008; LOMBARD; LECLERCQ, 2010; MOLDENHAUER, 2008; SORRIBES, 2008).

2.7.2.1 PETRIFILM™

The 3M Petrifilm system consists of a double film system: a base of polyethylene-coated squared paperboard, covered with a dehydrated culture medium containing a cold water-soluble gelling agent, nutrients and a removable transparent top film, which contains an indicator for activity and counting (FRANCO; LANDGRAF, 2005). The ingredients vary according to the plate, depending on the culture of the microorganisms. And a plating system called *all-in-one* (Figure 8). Instead of a Petri dish, the Petrifilm™ plate makes use of a thin plastic film that carries the culture medium (JASSON et al., 2010).

The Petrifilm™ system uses tetrazolium salts as an indicator, as it is an alternative indirect method for measuring respiratory activity associated with an electron transport chain. The reduction of the salt by the mitochondrial enzyme succinate dehydrogenase leads to the formation of an insoluble, intensely coloured precipitate known as Formazán. The tetrazolium salts: triphenyl tetrazolium chloride (TTC) and iodine-nitrophenyl tetrazolium chloride (INT) are often used because of the speed with which they are reduced by most dehydrogenase systems. The transformation only takes place in living cells and the amount of compound produced is proportional to the number of cells present (ORDONEZ, 2001).

The Association of Official Analytical Chemists (2002) recognises the Petrifilm ™ AC, CC, RCC, HSCC, EC, YM, RSA plate as an official method. The Association Française de Normalisation (AFNOR) also validates the Petrifilm plates AC, EB, CC, RCC, HSCC, EC, RSA as official (AFNOR VALIDATION, 2011). In Brazil, the Ministry of Agriculture, Livestock and Supply recognises the Petrifilm™ EC plate for animal products (BRASIL, 2005).

Figure 8 - Petrifilm TM system with dehydrated culture medium with gelling agent and removable transparent top film
Source: Cattani (2011)

a) PETRIFILM ™ for counting aerobes

The Petrifilm ™ plate for Aerobic Counting (AC) is a ready-made culture medium system that contains the nutrients of the standard counting agar, a gelling agent soluble in cold water and a tetrazole indicator to facilitate colony counting. As with conventional plates, the colony count range on Petrifilm™ plates is twenty-five to two hundred and fifty colonies, with all red colonies counted, regardless of their size or colour intensity (3M™ Company, St. Paul, MN, USA).

The samples to be tested must be prepared using appropriate diluents such as Butterfield's phosphate buffer, 0.1% peptone water, peptone saline diluent (ISO method 6887), buffered peptone water (ISO method 6579), saline solution (0.85-0.90%), letheen broth without sodium bisulphite and distilled water. The sample, diluted or not, is inoculated onto the surface of the base film in a volume of 1 ml and the top film is overlaid. Using a plastic diffuser, the sample is spread over a 20cm growth area^2 . After the gelling substance has solidified, the set is incubated at the temperature and for the time indicated by the manufacturer, according to the target microorganism. After incubation, the visible colonies are counted and the result is expressed in CFU/ml (ASSOCIATION OFFICIAL ANALYTICAL CHEMISTS, 2002).

In a study carried out to verify aerobic counts in ice cream, the results led to the conclusion that: the Petrifilm™ and Simplate® systems showed no significant difference between them, nor in relation to the other counting methods, and could be used as a technically viable alternative to the standard plate count in ice cream samples (SANT'ANA; CONCEIÇÃO; AZEREDO, 2002).

b) PETRIFILM ™ for counting Enterobacteriaceae

The Petrifilm ™ plate for counting *Enterobacteriaceae* (EB) is a ready-made culture medium system containing modified Violet Red Bile Glucose (VRBG) nutrients, a gelling agent soluble in cold water and a tetrazole indicator to facilitate colony counting. As with conventional plates, the colony counting range on Petrifilm ™ plates is fifteen to one hundred colonies, with all colonies producing acid and/or gas, red with yellow zones and/or red colonies with gas bubbles with or without yellow zones being counted (3M™ Company, St. Paul, MN, USA).

In studies carried out, the average count in log 10 and the estimated repeatability accuracy of the Petrifilm ™ EB method were similar to those of the VRBG method and better than those of the MPN method (SILBERNAGEL; LINDBERG, 2002; 2003).

c) PETRIFILM ™ for counting *Escherichia coli*

The Petrifilm ™ *Escherichia coli* Counting Plate (EC) is a ready-made culture medium system containing the nutrients of violet red bile agar (VRBA), a gelling agent soluble in cold water, a glucuronidase activity indicator (5-bromo-4-chloro-3indolyl-P-D-glycuronide) and a tetrazole indicator. Petrifilm ™ EC plates are used for counting *E. coli* and coliforms. The optimum colony count range on Petrifilm ™ EC plates is fifteen to one hundred and fifty colonies. Coliforms produce red colonies associated with gas bubbles and *Escherichia coli* produces blue colonies associated with gas bubbles. The glycoronidase produced by *Escherichia coli* reacts with the indicator dye on the plate, forming a blue precipitate around the colony. Non-coliform colonies are red, but are not associated with the gas bubbles (3M™ Company, St. Paul, MN, USA).

In a study carried out by Silva et al. (2006), Petrifilm® EC proved to be sensitive and efficient for detecting *E. coli* in food, when compared to the conventional multiple tube method, as well as having advantages such as: speed in obtaining results, practicality in carrying out and counting colonies and the possibility of complementary biochemical identification no more than 48 hours after incubation.

The correlations found between the counts of total coliforms and thermotolerant coliforms in the Compact dry® and Petrifilm ™ EC systems and the traditional methodology indicate that the results obtained by these rapid methods are equivalent to those obtained by the reference method, and that they can be used as viable alternatives in microbiological analyses of ground beef, without compromising reliability and sensitivity (CASAROTTI; PAULA; ROSSI, 2007).

The Petrifilm ™ EC was compared to the Australian method for enumerating coliforms and *E. coli* and no significant difference was found between the methods, except that the *E. coli* count was higher using the direct plating method with a resuscitation step than when counting using the Petrifilm ™ method (BLOCH et al., 1996).

2.7.2.2 TEMPO®

TEMPO® is a miniaturised system that arose from the concept of the microtiter plate (96 wells, 8x12 format), which allows for a reduction in the reactant volume and the medium to be used in the tests. It is based on the metabolism of specific substrates produced by micro-organisms and their detection using various indicator systems (SORRIBES, 2008). It was developed for the enumeration of microbiological quality indicators in food products (KUNICKA, 2007).

The test consists of a vial containing culture medium and a card, specific to the test (Figure 9). Dedicated equipment and specific programmes have been developed for the final reading of the sample. The process starts with the suspension of the sample to be tested, which is inoculated into the vial of culture medium. The inoculate is transferred to cards containing three series of sixteen wells (small, medium and large): 225 pl in the first well, 22.5 pl in the second well and 2.25 pl in the third, with a volume difference of 1 log between each series of wells. The card simulates the MPN method, using sixteen replicates instead of the three to five of the conventional method, reducing uncertainty and enabling accurate quantification. The card is then hermetically sealed, preventing any risk of contamination. The target microorganisms multiply in the culture medium, resulting in a signal detected by the equipment's TEMPO® Reader (based on the fluorescence of the indicator pH, 0-glucuronidase activity, etc.). Enumeration ranges from 10 to 49,000 CFU/ml or 100 to 490,000 CFU/ml, depending on the protocol (JASSON et al., 2010; OWEN; WILLIS; LAMPH, 2010).

The TEMPO® system offers significant savings by standardising analyses and minimising training time, the volume of waste and the number of operations. Workflow studies have shown a two- to three-fold reduction in sample handling time using TEMPO® compared to the ISO reference method (SOHIER; RANNOU; GORSE, 2006; CATTAPAN; VILLARD; DUMONT, 2009).

Figure 9 - Transferring the inoculum to the TEMPO® system cards
Source: Association of Official Analytical Chemists Research News (2009)

a) TEMPO® TVC

The TEMPO® TVC has been validated and certified by the AO AC *Research Institute* as an effective method for the total enumeration of aerobic bacteria in a variety of foods, including raw pork, raw beef, cooked and smoked meat products, among other products (CROWLEY et al, 2009; JOHNSON; MILLS; BEZZOLE, 2008).

The aerobic microorganisms present in the sample hydrolyse the substrate in the culture medium during incubation, producing a fluorescent signal by releasing the molecule 4-MU (methyl umbelliferone) into the medium (CROWLEY et al., 2009).

Several studies have been carried out to evaluate TEMPO® TVC, expressing a good correlation between it and the ISO reference method (GREEN et al., 2011; VERSETTI et al, 2011).

b) TEMPO® EB

The TEMPO® EB culture medium contains a fluorescent molecule, 4-methyl umbelliferone (4MU), which is a pH indicator (JOHNSON et al., 2008). When the pH is neutral, fluorescence is emitted. The *Enterobacteriaceae* present on the card assimilate nutrients from the culture medium during incubation, resulting in an increase in pH. When glucose is fermented by *Enterobacteriaceae*, the reagent is acidified, resulting in fluorescence extinction in tubes with a positive reaction (CÓLON-REVELES et al., 2007; JOHNSON; MILLS; BEZZOLE, 2009; OWEN; WILLIS; LAMPH, 2010).

In a study carried out on food and dairy products, no significant differences were found between TEMPO® EB, the NMP technique and plating for the enumeration of *Enterobacteriaceae,* demonstrating an advantage in the use of TEMPO® EB, as well as equivalence in the results found (OWEN; WILLIS; LAMPH, 2010). The method was validated by the AOAC after studies showed that it is statistically equivalent to the reference method (JOHNSON; MILLS; BEZZOLE, 2009).

c) TEMPO® EC

TEMPO® EC has the fluorogenic substrate 4-methylumbelliferyl-0-D-glucuronide (MUG) in the culture medium, which is cleaved by the enzyme 0-glucuronidase, produced by most strains of *E. coli*. The cleaved molecule, 4-MU, produces fluorescence (Figure 10), which is detected by the TEMPO® Reader (CROWLEY et. al., 2010).

In a study carried out by Torlak, Akan and Gokmen (2008), TEMPO® EC was clearly preferable to TBX medium, specifically for samples that include low levels of contamination and cold-stressed cells, proving to be a suitable alternative method to standard plate counting for the enumeration of *E. coli*, showing a high correlation coefficient in the results found.In studies on the enumeration of *E. coli,* described in the validation process by the AOAC *Research Institute,* the TEMPO® EC method proved to be statistically equivalent to the reference method (JOHNSON, 2007).

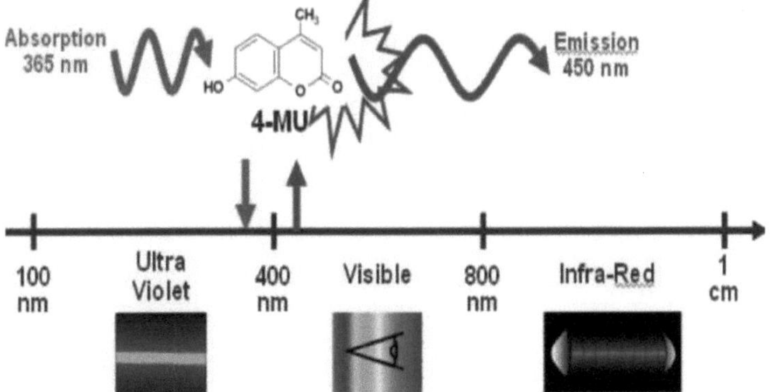

Figure 10 - Fluorescence emitted after MUG cleavage in the TEMPO® System
Source: BioMerieux (2008)

2.7.23 E.COLI LIVES 0157 (ECO LIVES)

The VIDAS® System (bioMerieux, Marcy-I'Étoile, France) is a system based on the ELISA technique, which detects micro-organisms such as *Salmonella, Campylobacter, Listeria monocytogenes, Escherichia coli* 0157, among others. The enzyme-linked immunosorbent assay (ELISA) is a detection system that uses monoclonal antibodies on plates to capture the target antigen. The captured antigen is then detected using a second antibody that may be conjugated to an enzyme. The addition of a fluorescent substrate makes it easier to visualise the target antigen. This method offers specificity and potential automation (FORSYTHE, 2002; REITER et al., 2010).

VIDAS® *E. coli* 0157 (VIDAS® ECO) is a qualitative test, automated on VIDAS® systems, which allows the detection of *Escherichia coli* serotype 0157 in human food products using the ELFA (Enzyme Linked Fluorescent Assay) technique. The system can analyse twelve samples simultaneously after the first test, when the equipment is calibrated using a calibrator identified as SI that will be analysed in duplicate, a positive control (Cl) and a negative control (C2). The calibrator must fall within the RFV ("Relative Fluorescence Value") limits set (bioMerieuex, Marcy-l'Étoile, France). For ELISA, the typical detection limit is 10^5 cells/ml (BLACKBURN; MCCARTHY, 2000).

The cap used in the system consists of ten wells covered with a sealed and labelled aluminium foil, which contains a barcode with information on the test code, batch number and expiry date of the package. In VIDAS® ECO, the single-use cone (SPR®) serves as both the solid phase and the pipetting support. The inside of the cone is coated with a recombinant phage protein that enables the capture of *E. coli* O157:H7 (BIOMERIEUX, 2008).

All the test steps are carried out automatically by the device and consist of a succession of cycles of aspiration and dispersion of the reaction medium. The *E. coli* O157:H7 present are captured by the recombinant phage protein fixed inside the cone. The elements that remain free are eliminated by washing. The alkaline phosphatase conjugate is then aspirated through the cone and attaches to the *E. coli* O157:H7, which are already attached to the phage protein on the cone wall. Further washing steps remove the unfixed conjugate. In the final revealing step, the substrate 4-methylumbelliferylphosphate is aspirated and dispensed into the cone; the enzyme in the conjugate catalyses the hydrolysis reaction of this substrate into 4-methylumbelliferone, whose emitted fluorescence intensity is measured using an optical scanner at 450 nm in VIDAS® and the value is expressed as RFV (Relative Fluorescence Value). Once the test is complete, the results are automatically analysed by the device, which provides a test value for each sample. This value is compared with the internal references (thresholds) and each result is interpreted (positive, negative). The values are interpreted and expressed as follows: test value < 10 = negative interpretation; test value >10= positive interpretation. A result with a test value lower than the threshold value indicates a sample that does not contain 0157 antigens or that contains a concentration of 0157 antigens lower than the detection limit. A result with a test value greater than or equal to the threshold value indicates a sample contaminated with *E. coli* 0157. In this case, the results should be confirmed by the conventional method in Mac Conkey Broth with cefixime and potassium tellurite (CT-MAC). Sorbitol Mac Conkey Agar with cefixime and potassium tellurite uses sorbitol instead of lactose and is more selective with a lower concentration of bile salts and crystal violet for inhibiting Gram-positive bacteria, mainly *Enterococcus and Staphylococcus*, as well as *Proteus* species (BIOMERIEUX, 2008).

When performing the VIDAS ECO test, cross-reactions can occur between the test antibodies and strains/cepas that have antigen 0157 other than *E. coli*, such as some *Salmonella* of group N, some *Citrobacter, M. morganii and E. cloacae* (DE BOER; HEUVELINK, 2000; HUANG et al, 2005).

Huang et al. (2005) and Reiter et al. (2010) carried out comparative studies between the VIDAS® system and other alternative methods in relation to the traditional method, with different microorganisms and in different food samples. The VIDAS® system was found to have better diagnostic performance and effectiveness compared to the other methods tested.

The VIDAS ECO test was validated by the Association Française de Normalisation (AFNOR) - BIO 12/08 - 07/00, on 5 July 2000, for all food products; by AOAC RI Performance Test Method No. 010504 for minced beef and by the Chinese Government SN/T 0184.1 (AFNOR, 2011; ASSOCIATION OFFICIAL ANALYTICAL CHEMISTS, 2002). The antigen-antibody reaction in the VIDAS® ECO system can be seen in Figure 11.

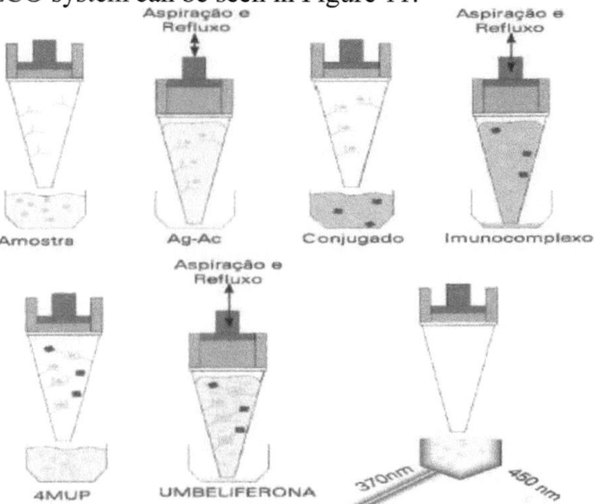

Figure 11 - Capture of the antigen present in the sample; Ac binds to Ag; a second antibody conjugated to an enzyme forms the immunocomplex; The addition of a fluorescent substrate 4MUP facilitates visualisation of the target antigen and formation of umbelliferone; read d

Source: BioMerieux (2008)

CHAPTER 3

EVALUATION OF THE INFLUENCE OF LACTIC ACID AND HOT WATER ON MICROBIAL DECONTAMINATION OF PORK CARCASSES IN BRAZIL

In order to improve industrial health conditions and to increase microbiological safety and hygiene in the slaughter process, it became compulsory in 1997 for slaughterhouses operating under the rules of the Federal Inspection Service in Brazil for the production of meat and animal products to implement Good Manufacturing Practices (GMP) programmes (BRASIL, 1997) and Hazard Analysis and Critical Control Points (HACCP) programmes (BRASIL, 1998).

However, the sanitary condition of the pig before slaughter, especially in relation to *Salmonella* spp, is a factor that contributes to bacterial contamination of the carcass during health inspection procedures, due to the cuts made to the lymph nodes and concomitant manipulation of the abdominal organs and viscera, leading to cross-contamination of the carcasses at slaughter (BESSA; COSTA; CARDOSO, 2004; HUMPHREY; JORGENSEN, 2006; OLIVEIRA et al., 2010; SILVA et al., 2009). It is therefore necessary to adopt control measures to prevent the spread of pathogens present in slaughtered animals to humans (NESBAKKEN; SKJERVE, 1996; SOFOS; BELK; SMITH, 1999).

Bacterial decontamination methods on carcasses may be necessary as an effective intervention to reduce contamination after slaughter (SWANENBURG et al., 2001). These decontamination processes based on immersion, washing or spraying with water or chemical solutions, washing with hot water at temperatures above 74°C, pasteurisation with steam, vacuuming with steam, chemical washing with organic acid solutions (HUFFMAN, 2002; KOOHMARAIE et al., 2005) are used in several countries, such as the USA, Canada and Australia (SOFOS; BELK; SMITH, 1999). In this way, carcass decontamination is a relevant option in cases where the prevalence of pathogens in carcasses must be reduced to levels that cannot be achieved by implementing hygiene practices at the slaughterhouse or by interventions at primary production level (LAWSON et al., 2009).

The aim of this study was to assess the influence of the use of lactic acid and washing water on the microbial population naturally present on the surface of pig carcasses in the state of Santa Catarina, Brazil.

3.1 MATERIAL AND METHODS

From March 2010 to March 2011, 152 samples were analysed from 76 pig carcasses collected at the end of the line at a slaughterhouse under Federal Inspection in the state of Santa Catarina, where 4,500 pigs are slaughtered every day.

3.1.1 Samples

Each sample consisted of sponge swabs collected from four points on the carcass (shank, belly, loin and jowl). The four sponge swabs together (100 cm² each) represent the sampling of an area corresponding to 400 cm² . The samples from the carcass surfaces were collected by pressing the sponge, previously moistened in 100 mL of sterile 0.1% peptone saline solution, onto sterilised delimiting moulds in the area to be sampled (100 cm^2) for around 20 seconds, starting the procedure vertically, then rubbing horizontally and finally diagonally. The sponge was placed in a sterile stomacher bag and kept refrigerated at 4 °C until it was analysed in the laboratory, which was no more than 24 hours after collection.

The samples were collected before and 12 hours after the following treatments were applied: treatment zero (TO), water sprayed at room temperature; treatment 1 (Tl), water sprayed at between 76°C and 78°C, treatment 2 (T2), water sprayed at between 76°C and 78°C followed by lactic acid spray at a concentration of 1.5%; treatment 3 (T3), water sprayed at room temperature followed by lactic acid spray at a concentration of 1.5%. The carcasses were refrigerated at the time of collection 12 hours after the treatments.

3.1.2 Methodology used for sprinkling water and lactic acid

To spray water on the surface of the carcasses, the final shower on the slaughter line was used, with dimensions of 3 metres long, 3 metres high and 1.15 metres wide inside; and 67 cm at the shower entrance. The water in the shower is automatically activated by a sensor for 5 seconds, and is sprayed by means of three rows of sprinkler nozzles (14 nozzles in the first row, 20 nozzles in the second row and 10 nozzles in the third row), totalling 44 sprinkler nozzles, under a pressure of 2.2 Kgf/cm^2 .

The water used was between 14°C and 25°C for TO and T3. The heated water used in the experiment for TI and T2 was heated to 86°C inside the water tank and reached the sprinkler nozzles of the final shower between 76°C and 78°C.

The 1.5% lactic acid solution at room temperature was sprayed using a hand-held backpack sprayer. The sprinkler nozzle maintained a constant outlet pressure of around 2 Kgf/cm and a flow rate of 300 ml/minute.

Natural lactic acid L(+) produced by fermenting sugar cane was used, with a concentration of 84.5-85.5%, brand name PURAC®. 150 mL of lactic acid was used, added to enough water for 10 litres of solution, and the pH was kept at around 2.5, controlled with a pH meter. The carcasses were sprayed manually for an average of 3 minutes, with a volume of approximately 900 mL of solution per carcass, forming a uniform film of solution on the surface of the carcass. The carcasses that received this treatment were sprayed immediately after the final shower on the slaughter line, located before entering the heat shock and refrigeration chambers.

The carcasses were subjected to sensory analysis after reaching a temperature of 7°C.

3.1.3 Methodology used for the microbiological tests

The microbiological tests to analyse mesophilic aerobic microorganisms, enterobacteria, *E. coli, Salmonella* spp. and *E. coli* 0157 were carried out according to the methodology described in the American Public Health Association (APHA)

(ANDREWS et al., 2001; KORNACKI; JOHNSON, 2001; MENG; ZHAO; DOYLE, 2001).

The *Salmonella* spp. strains were typed using the Polymerase Chain Reaction (PCR) technique at the Enterobacteria Laboratory of the Oswaldo Cruz Foundation (FIOCRUZ, RJ).

Samples of the water used to wash the carcasses were collected on each day of the experiment and analysed for total counts of mesophilic aerobic microorganisms, enterobacteria, *E. coli* and *Salmonella* spp.

3.1.4 Statistical Analysis

The data on the micro-organisms surveyed was statistically analysed using STATISTICA 7.0, with analysis of variance at a 95% confidence level (ANDRADE; OGLIARI, 2007). The counts in CFU/cm^2 of mesophilic aerobic micro-organisms, enterobacteria and *E. coli* were transformed into logs for reduction. Negative log values express counts less than or equal to zero.

3.   2RESULTS

Table 1 shows the results of the microbiological tests carried out on *Salmonella* spp., *E. coli* and enterobacteria in the pig carcasses before and after the treatments.

Table 1 - Prevalence of pig carcasses detected positive for *Salmonella* spp., *E. coli* and Enterobacteriaceae, analysed before and after the application of water and 1.5% lactic acid treatments (TO,T1,T2,T3)

Micro-organism	Phase Treatment	Number of positive carcasses				
		TO	IT	T2	T3	Total
Salmonella	Before	1	1	0	1	3/76
	Then	0	1	1	0	2/76
E. coli	Before	8	8	5	10	31/76
	Then	5	5	0	3	13/76
Enterobacteriaceae	Before	11	15	9	9	44/76
	Then	8	8	0	8	24/76

Source: Author's data

Note:T0, water at room temperature ; Tl, water at a temperature between 76°C and 78°C ; T2, water at a temperature between 76°C and 78°C followed by spraying with lactic acid at a concentration of 1.5% ; T3, water at room temperature followed by spraying with lactic acid at a concentration of 1.5%.

Salmonella spp. was detected in 5 of the 76 carcasses sampled, representing a prevalence of 6.58% of the sample. The *Salmonella* spp. strains were identified and belonged to three serovars, with *Salmonella* Thyphimurium (2/5), *Salmonella* Agona (2/5) and *Salmonella* Derby (1/5) standing out. It was not possible to statistically evaluate the effect of the treatments on the decontamination of carcasses naturally contaminated with *Salmonella* spp. due to the low frequency of carcasses found with the pathogen.

The frequency of *Salmonella* spp. found before treatment differed little from the frequency found after treatment (1.02%) and

the prevalence of *E. coli* in the carcasses was reduced by 32.68% after the treatments and for enterobacteria the reduction after the treatments was 26.31%.

E. coli counts ranged from $0.1.1o^1$ CFU/cm^2 to $9.2.1o^1$ CFU/cm^2 before treatments and from $0.1.1o^1$ CFU/cm^2 to $0.9.1o^1$ CFU/cm after treatments, and enterobacteria counts ranged from $0.1.1o^1$ CFU/cm^2 to $1.3.10^3$ CFU/cm^2 before the treatments and from $0.1.1o^1$ CFU/cm^2 to $2.2.10^2$ CFU/cm^2 after the treatments. The treatment that showed a 100 per cent reduction in the number of carcasses positive for *E.coli* and enterobacteria was treatment 2, which used water with a temperature between 76 °C and 78 °C followed by spraying with lactic acid at a concentration of 1.5 per cent.

The differences between the results of the initial and final counts of mesophiles, enterobacteria and *E. coli* after the treatments were analysed. Table 2 shows the average values found in the initial (before treatment) and final (after treatment) counts for each microorganism investigated, and their reduction after the treatments.

Table 2 - Average count of aerobic mesophilic microorganisms, enterobacteria, *E. coli*, initial (before treatment), final (after treatment) and the reduction achieved after applying the treatments to decontaminate pig carcasses.

Micro-organisms	Stage of operation	Treatments			
		T0	T1	T2	T3
		Average	Average	Average	Average
Mesophiles	Before	$3,68^a$	$3,91^a$	$3,79^a$	$3,79^a$
	Then	$3,05^b$	$3,01^b$	$2,92^b$	$3,21^b$
	Reduction	0,67	0,96	0,92	0,54
E. coli	Before	$0,48^c$	$0,43^c$	$0,02^c$	$0,30^c$
	Then	$0,19^d$	$0,01^d$	$-0,04^d$	$-0,01^d$
	Reduction	0,33	0,47	0,07	0,31
Enterobacteriaceae	Before	$0,59^e$	$0,76^e$	$0,28^e$	$0,30^e$
	Then	$0,49^f$	$0,22^f$	$-0,04^f$	$0,23^f$
	Reduction	0,09	0,54	0,32	0,07

Source: Author's data

Note: water at room temperature (T0); water with a temperature between 76°C and 78°C (Tl); water with a temperature between 76°C and 78°C followed by spraying with lactic acid at a concentration of 1.5% (T2), water at room temperature followed by spraying with lactic acid at a concentration of 1.5% (T3).

The greatest reduction obtained for aerobic mesophilic micro-organisms was with treatment 1 (water with a temperature between 76°C and 78°C) which reduced the total count of mesophilic micro-organisms by 0.96 log CFU/cm^2 , followed by treatment 2 (water with a temperature between 76°C and 78°C followed by spraying with lactic acid at a concentration of 1.5%) which reduced the count by 0.92 log CFU/cm^2 , but statistically there was no significant difference ($p < 0.05$) in the reduction of micro-organisms with the treatments tested. All the carcasses sampled were prevalent before and after the treatments for mesophilic micro-organisms, with counts ranging from 3.5. 10 CFU/cm^2 to 1.0. 10^5 CFU/cm^2 before the treatments and 1.8. 10^1 CFU/cm^2 up to $1.8.10^4$ CFU/cm^2 after the treatments.

There were no positive samples for *E. coli* 0157, so it was not possible to assess the effect of the treatments on this microorganism.

All the results of the analyses of the water used in the study were negative for *Salmonella* spp. and showed counts of less than 1 UFC/mL for mesophilic microorganisms, enterobacteria and *E. coli*.

In the sensory test, there was no colour change in the skin of the carcasses in the different treatments applied.

There was no statistical difference at a significance level of 95% for any of the treatments used in this experiment with regard to microbial reduction in logarithmic cycles.

3.3   DISCUSSION

In this study, the effectiveness of hot water, 1.5% lactic acid and hot water followed by 1.5% lactic acid spraying for decontaminating pork carcasses was compared. The treatments were applied using the structure of the processing plant, without altering the flow or equipment of the slaughterhouse. As there are no studies on the use of decontaminants on pork carcasses in Brazil, it is difficult to discuss the results found, and some discussions refer to studies carried out on pork cuts or beef carcasses.

The reduction of mesophilic microorganisms naturally present on the surface of pork carcasses, verified in this work, was approximately one logarithmic cycle, after IT and T2, and was below the reduction found by Reagan et al. (1996) and by Gill et al. (1995) of two logarithmic cycles. Reagan et al. (1996) concluded that hot water from 74°C to 87.8°C (for 18 seconds and a pressure of

2.4kPa) applied to the surface of artificially contaminated bovine carcasses was effective in reducing contamination by 2 logarithmic cycles of mesophiles and significantly reduced the incidence of *Salmonella* spp. However, aspects such as exposure time and sprinkler nozzle pressure must be taken into account, as they have a substantial impact on the magnitude of microbial reduction.

Similarly, Gill et al. (1995), in their experiment to decontaminate pigs with hot water at various temperatures in a commercial establishment, concluded that spraying water at a temperature of 85°C for 20 seconds reduced the total bacterial count by two logarithmic cycles. Gill and Landers (2003), studying various carcass decontamination treatments, concluded that spraying carcasses with 2% lactic acid was not as effective as spraying water at a temperature of 85°C for 10 seconds and with a nozzle pressure of $20kg/cm^2$.

In a study by Lawson et al. (2009), water treatment at 80°C for 15 seconds was the most effective method for reducing *Salmonella* spp. and *Yersinia* spp. from the surface of pork carcasses. In our study, the exposure time of 5 seconds used in the final carcass washing shower, and the pressure of the sprinkler nozzles, which was around 2.2 Kgf/cm^2 (below the 3 Kgf/cm^2 recommended by Brazilian legislation), proved to be ineffective in reducing and physically removing micro-organisms.

Treatment 1 (Tl), which used hot water and then refrigerated the carcass, reduced the total count of mesophilic microorganisms by around 1 log CFU/cm and enterobacteria and *E. coli* by 0.5 log CFU/cm^2, while the control treatment (T0), which used only water at room temperature, also followed by refrigeration of the carcass, did not achieve the same reduction. Refrigeration alone (data for T0) was shown to reduce mesophilic aerobic microorganisms by 0.67 log and *E. coli* and enterobacteria by 0.33 and 0.05 log CFU/cm, respectively. This was similar to that obtained in treatment 3 (T3), which used a water bath at room temperature followed by 1.5% lactic acid spray, and the carcass was then refrigerated. The data presented in this study is similar to the reduction of around 1 log obtained with the application of hot water with a temperature >82°C by Gill and Bryant (1997). The low microbial reduction found in T0 corroborates the ICMSF, 2005, which states that only bathing the carcasses and then refrigerating them has little effect on microbial reduction. For hot water to be effective in microbial reduction, it must be maintained at a temperature >74°C on the surface of the carcass for longer than 5 seconds (KOOHMARAIE et al., 2005).

With regard to treatment 3 (T3), which used a bath with water at room temperature followed by the sprinkling of 1.5% lactic acid, there was little microbiological reduction (0.54log UFC/cm^2 of mesophiles and 0.31log UFC/cm of *E. coli*). The low level of reduction achieved by lactic acid compared to hot water was demonstrated in the study carried out by Bosilevac et al. (2006), who used 2% lactic acid, which was not as effective at reducing microorganisms (1.6 log CFU/cm) as hot water at 76°C for 5.5 seconds, which reduced 2.7 log CFU/cm^2 of mesophilic microorganisms and enterobacteria in beef carcasses. According to studies carried out by the International Commission on Microbiological Specifications for Foods (2005), applying 1% lactic acid to the surface of beef and pork carcasses reduces the count of mesophilic microorganisms by 0.8 to 1.9 log, which is closer to the data found in our study.

Lactic acid is widely used in slaughterhouses in the United States. In some, it is used as part of GMP programmes due to its lethal and injurious effect on bacterial cells. Some requirements have been listed for its effect to be effective as a decontamination method: acid concentration between 2.5 and 10% (v/v) to allow the effect when diluted and applied to the carcasses; acid pH 2.8; acid temperature 25-55°C; application pressure 13.8 - 27.6 Kgf/cm; volume of acid applied 500 ml; and 35 seconds of application. (BOLTON; DOHERTY; SHERIDAN, 2001). In our study, given that there were no changes to the equipment used in the slaughter routine, we met almost all the requirements with the exception of the concentration used, which was lower (1.5%) and the pressure used, which was 2 Kgf/cm^2, which may explain the low microbial reduction achieved when it was sprayed on the carcasses.

There was no visual change in the colour of the skin of pork carcasses treated with hot water and 1.5% lactic acid, in agreement with the study by Pipek et al. (2005) that there is no variation in the appearance of the surface of pork carcasses treated with 2% lactic acid and steam (pasteurisation at 82°-97°C).

In a study carried out by Epling, Carpenter and Blankenship (1993), a reduction in *Salmonella* spp. and *Campylobacter* spp. was observed in pig carcasses after sprinkling 2% lactic acid, which cannot be observed in this study, as the prevalence was low in the carcasses analysed and there was no homogeneous distribution of the frequency of *Salmonella* spp. in the carcasses subjected to the different treatments so that the reduction could be assessed.

It can be seen in this study that *Salmonella* spp. was present in the pig carcasses before and after the treatments, demonstrating the permanence of the pathogen through cross-contamination by equipment or handlers during slaughter, or through direct contact between the carcasses during the refrigeration process. The same conclusion was reached in a study carried out by Swanenburg et al. (2001) on the slaughter of known seropositive and seronegative pig herds, where cross-contamination between carcasses occurred during the slaughter and refrigeration procedures.

Brazil does not have an official programme for the identification and control of *Salmonella* spp. in pig herds, so it is not known whether or not the herds entering slaughter are seropositive for *Salmonella* spp.

In our study, the prevalence of *Salmonella* spp. was 6.58% (5/76), similar to that found in Denmark, where 30 different serotypes of *Salmonella enterica* were isolated from 832 pigs (6.2%) in the slaughter of 13468 finishing pigs, with *Salmonella* Typhimurium being the predominant serotype in 536 (64.4%) of the isolates (BAGGESEN et al., 1996).

The serotypes Typhimurium (2/5), Agona (2/5) and Derby (1/5) found in this study, which was carried out in Santa Catarina, Brazil, are similar to the study carried out on pig carcasses by Seixas, Tochetto and Ferraz (2009) and are similar to the serotypes present in the pig herd of Santa Catarina, Brazil, demonstrated in the work carried out by Kich et al. (2011), where the serotype Typhimurium was the most commonly found. Similarly, Bessa, Costa and Cardoso (2004) reported the prevalence of the Typhimurium, Agona and Derby serotypes in pigs slaughtered in southern Brazil.

By comparing the studies of various authors in relation to the exposure time and pressure used in the final bath of the pork carcass with the time and pressure of the shower used in the slaughter process in Brazil, which does not use decontamination methods, we can see the need to increase the exposure time of the pork carcasses in the final bath of the slaughter and increase the water outlet pressure of the sprinkler nozzles in order to improve the microbial reduction process.

3.4 CONCLUSIONS

Although there was no statistically significant difference in the reduction of contamination by micro-organisms with the treatments used in this study, the reduction can be observed.

The treatment that proved 100% effective in reducing the number of carcasses contaminated with enterobacteria and *E. coli* was treatment 2, which consisted of applying water at a temperature of between 76°C and 78°C followed by spraying with lactic acid at a concentration of 1.5%.

Using water with a temperature above 76°C for the final bath of pork carcasses, before the carcasses enter the cold rooms, has been shown to be more efficient for microbial reduction than water at room temperature as it is used in the slaughter process in Brazil.

Considering the 6.58% prevalence of *Salmonella* spp. found in this study, and since it is a pathogen of public health importance, it is necessary to consider adopting decontamination practices in order to minimise the spread of the pathogen through pig carcasses to the consumer. This includes adopting a programme to identify batches that are positive for *Salmonella* spp. before the animals enter the slaughterhouse.

In this way, you can separate positive batches from negative batches and establish a slaughter order for these animals to minimise or prevent contamination.

CHAPTER 4

ALTERNATIVE METHODS FOR COUNTING MICROORGANISMS IN PIG CARCASSES

The presence of micro-organisms in pork is the result of contamination of live animals, equipment, handlers and the processing environment (NESBAKKEN; SKJERVE, 1996). This demonstrates the importance of managing the risks associated with meat safety, developing strategies to control pathogens at all stages of processing (SOFOS; GEORNARAS, 2010) and the need to use methodologies for detecting micro-organisms quickly and efficiently in routine diagnostics in order to provide adequate information on the possible presence of pathogens in raw materials and finished products to control the production process and monitor industrial hygiene and cleaning practices (BOER; BEUMER, 1999).

Microbiological detection methods are often categorised into two groups: conventional or traditional and rapid or alternative. Traditional methods are characterised by being laborious, employing large volumes of culture media and requiring considerable time to obtain analyses and results (JASSON et al., 2010). Rapid methods are an alternative to conventional methods and are designed to obtain the final result in less time. This is highly desirable in the food industry, even though the techniques are more expensive and require highly trained personnel (FORSYTHE, 2002).

Considering that Brazil is a major exporter of pork and the rapid need to know the microbiological quality of this product, together with the lack of research into this matrix, this study aimed to verify the time required to carry out the analyses and the correlation between the conventional methodology and the alternative methodologies of the Petrifilm ™ system (3M) and Tempo® system (bioMeriéux) for the diagnosis of *Escherichia coli,* enterobacteria and mesophilic microorganisms in pork matrix.

4.1   MATERIAL AND METHODS

During the period from March 2010 to March 2011, 152 pig carcasses were analysed, collected at the end of the line of a slaughterhouse under Federal Inspection and with a daily slaughter of 4,500 pigs. Each sample consisted of sponge *swabs* taken from four points on the carcass (shank, belly, loin and jowl) and packed in a single sterile *stomacher* bag. The four sponge *swabs* together (100 cm each) represent the sampling of an area corresponding to 400 cm^2 . The samples from the carcass surfaces were collected by pressing the sponge, previously moistened in 100 mL of sterile 0.1% peptone saline solution, onto sterile delimiter moulds in the area to be sampled (100 cm^2) for around 20 seconds, starting the procedure vertically, then rubbing horizontally and finally diagonally. The sponge was placed in a sterile *stomacher* bag and kept refrigerated at 4 °C until it was analysed in the laboratory, which was no more than 24 hours after collection. To prepare the sample used for all the methodologies, the *stomacher* bag containing the sponge was first rubbed in 0.1% peptone water and from this homogenate, dilutions of 1:10 and 1:100 were prepared (BRASIL, 2007).

In the conventional methodology, the homogenised sample diluted 1:10 and 1:100 was inoculated onto agar (Oxóid) specific for each type of micro-organism investigated, following the recommendations for the conventional methodology for mesophilic micro-organisms, enterobacteria and *E. coli* (KORNACKI; JOHNSON, 2001; MORTON, 2001).

The alternative Petrifilm™ methodology is a so-called *all-in-one* plating system in which the Petrifilm™ plate makes use of a thin plastic film that carries the culture medium (JASSON et al., 2010). The ingredients of this culture medium vary for each type of microorganism to be identified. To count mesophilic microorganisms in this system, 1 mL of the homogenate and the 1:10 and 1:100 dilutions were inoculated onto Petrifilm™ AC cards and incubated at 36 °C ± 1 °C for 48 hours. To count enterobacteria, 1 mL of the homogenate and the 1:10 and 1:100 dilutions were inoculated onto Petrifilm™ EB cards and incubated at 36 °C ± 1 °C for 24 hours. For the *Escherichia coli* count, 1 mL of the homogenate and the 1:10 and 1:100 dilutions were inoculated onto Petrifilm EC cards and incubated at 36 °C ± 1 °C for 24 hours. After incubation, the colonies were counted and the result was recorded in CFU/cm² (KORNACKI; JOHNSON, 2001).

In the methodology used for the Tempo® system for the total count of mesophilic microorganisms (TVC), the TVC medium was reconstituted with 3.0 mL of sterilised distilled water. Then 1 mL of the homogenised sample prepared without dilution was added, making a total volume of 4 mL. For the enterobacteriaceae (EB) count, the EB medium was reconstituted with 3.9 mL of sterilised distilled water, adding 0.1 mL of the homogenised sample prepared without dilution. For the *Escherichia coli* (EC) count, the EC medium was reconstituted with 3.9 mL of sterilised distilled water and then 0.1 mL of the homogenised sample was added. The vials and cards were placed inside the equipment, where the contents of the vial were transferred to the card containing 48 wells with three different volumes (16 x 225 pL; 16 x 22.5 |1L; 16 x 2.25pL). Immediately after filling, the cards were sealed. This operation took place inside the Tempo® preparation workstation. The cards were incubated in an oven for 24 hours at 35°C for EB and EC and for 48 hours at 35°C for TVC. After this period, the cards were read at the Tempo® reader station. Once the reading was complete, the results were analysed automatically by the system, which determined which wells were positive, generating the results in CFU/cm² .

Linear regression analysis was used to statistically analyse the data, determining Pearson's linear correlation coefficient (r), which ranges from -1 to 1, to quantify the correlation between the methodologies used. Within this range, the values were distributed as follows: 0 for no correlation, 0.30 to 0.70 for moderate correlation and 0.70 to 1 for strong correlation, with no negative correlation results (-1). The data relating to the micro-organisms investigated was statistically analysed using STATISTICA 7.0, with analysis of variance and Tukey's test at a 95% confidence level (ANDRADE; OGLIARI, 2007).

4.2   RESULTS AND DISCUSSION

In this study, the reduction in execution time between the methodologies used can be observed. For enterobacteria, the reduction in time from sample preparation to the final result issued by the Petrifilm™ system and the Tempo® system was 24 hours compared to the conventional methodology. For the *E. coli* count, there was a reduction of 144 hours for the two alternative systems compared to the conventional methodology.

The results of the mesophilic microorganism, enterobacteria and *E. coli* counts show that there was agreement between the different methodologies used, when examining the correlation coefficient (r), which shows a positive linear association between the methodologies, as shown in Table 3, in agreement with studies presented by Crowley et al. (2009), Sohier et al. (2007) and Owen, Willis and Lamph (2010).

Table 3 - Correlation coefficient (r) between the different methodologies (Conventional, PetrifilmTM and Tempo®) for counting mesophilic microorganisms, enterobacteria and E. coli on the surface of pig carcasses from March 2010 to March 2011

Methodologies	Mesophiles	Enterobacteriaceae	*Escherichia coli*
Conventional x Tempo®	r= 0,6712	r=0,7535	r=0,6255
Conventional x Petrifilm™	r=0,7770	r=0,8368	r=0,3955
Petrifilm™ x Time®	r=0,8118	r=0,7480	r=0,5379

Source: Author's data

Note: r = Pearson's correlation coefficient.

With regard to mesophilic microorganisms, it can be seen that these were detected in all the methodologies used. The average count results were similar between the methodologies, with no statistical difference between the means, with 3.44 log of mesophilic microorganisms being found for the conventional method, 3.47 log of mesophilic microorganisms for Petrifilm™ and 3.34 log of mesophilic microorganisms for Tempo®, with a standard deviation of 0.66; 0.63 and 0.66 respectively.

Escherichia coli was detected in all the methods used. The presence of *E. coli* in Tempo® was 32.89% (50/152), in Petrifilm™ was 38.82% (59/152) and in the conventional method was 31.58% (48/152). Enterobacteria were detected in all methods, 44.08% (67/152) in Tempo® , 61.84% (94/152) in Petrifilm™ , and 19.08% (29/152) in the conventional method. There was no significant difference between the means of mesophilic microorganisms, *E. coli* and enterobacteria in the methods tested.

Cattapan et al. (2007), Johnson (2007), Kunicka (2007) and Sohier et al. (2007) verified the reduction in time in workflow studies that demonstrated a two- to three-fold reduction in sample handling time with the use of Tempo® compared to the ISO reference method. These authors report that the reduction and optimisation of sample testing time is due to the type of system used by the fully automated Tempo® equipment, which is based on the metabolism of specific substrates produced by micro-organisms. This miniaturised system uses the concept of the micro-titre plate, which reduces the reactant volume and the medium to be used in the tests. The time optimisation and specificity of the test using the Tempo® system methodology can also be seen in the results of this study.

The Petrifilm™ system's reduction in test time and response is due to its simplicity of handling and speed of counting. The process of identifying microorganisms is carried out using tetrazolium salts as indicators. This process is considered an alternative indirect method for measuring the respiratory activity of microorganisms. The reduction of this salt by the mitochondrial enzyme succinate dehydrogenase leads to the formation of an insoluble, intensely coloured precipitate. The tetrazolium salts used, triphenyl tetrazolium chloride (TTC) and iodine-nitrophenyl tetrazolium chloride (INT), are rapidly reduced by most microorganism dehydrogenase systems. The transformation only takes place in living cells and the amount of compound produced is proportional to the number of cells present. Silva, Cavalli and Oliveira (2006) recommend the use of Petrifilm™EC as a sensitive and efficient method for counting *E. coli* in food of animal origin instead of the conventional method. The simplicity and speed of this methodology could be verified in the tests carried out in this study using the Petrifilm™ system.

There was a moderate correlation between the conventional methodology and Tempo® (i=0.62) for *Escherichia coli*. The results also show a moderate correlation between the conventional methodology and Petrifilm ™ (r= 0.39) and between the two alternative methodologies Tempo® and Petrifilm ™ (r= 0.53). The agreement between the conventional methodology and Tempo® for detecting *Escherichia coli* in this study was lower than the agreement found in the tests carried out by Green et al. (2007), with i=0.92 between the Tempo® method and the conventional plate methodology for assessing *Escherichia coli* in samples of beef and poultry products. The correlation was also lower than that found in the comparative studies carried out by Devulder et al. (2007), r= 0.99 in meat products. We can attribute the differences in correlation between the different studies in the detection of *Escherichia coli* to the microbiological quality of the samples.

Studies by Sohier et al. (2007) show similar or higher levels of repeatability and reproducibility with Tempo® compared to the standard plate counting method.
Crowley et al. (2010) show in their work that for the detection of *E. coli*, Tempo® EC contains the fluorogenic substrate 4-methylumbelliferyl-0-D-glucuronide (MUG) in the culture medium, which is cleaved by the enzyme 0-glucoronidase, produced by most strains of *E. coli*. When cleaved, the MUG produces fluorescence which is then detected by the Tempo® Reader. Among the methods used in this study, the best agreement for detecting *E. coli* (r= 0.62) was found between the conventional methodology and Tempo® , which demonstrates the method's sensitivity and specificity for detection, confirmed by the results found in this study.

For enterobacteria, there was a strong correlation between the conventional methodology and Tempo® (r= 0.75), between the conventional methodology and Petrifilm ™ (r= 0.83), and between the two alternative methodologies Tempo® and Petrifilm ™ (r= 0.75). A study by Owen, Willis and Lamph (2010) found correlation coefficients equivalent (r= 0.75 to 0.78) to those found in this study, when evaluating Tempo® EB in relation to the most probable number technique and surface and deep seeding. The detection of enterobacteria by Tempo® was demonstrated in a study carried out by Sohier et al. (2007), with similar or higher levels of repeatability and reproducibility of this method when compared to the standard plate count method.

The specificity of Tempo® EB for detecting enterobacteria can be understood by the reaction that occurs between the microorganism and the culture medium present in the vials, which contains a fluorescent molecule, 4-methyl umbelliferone (4MU), which is a pH indicator (JOHNSON; MILLS; BEZZOLE, 2008). When the pH is neutral, fluorescence is emitted. The microorganisms of the *Enterobacteriaceae* family present in the cardboard assimilate the nutrients in the culture medium during incubation, resulting in an increase in pH. When glucose is fermented by these microorganisms, the reagent is acidified, resulting in fluorescence extinction in tubes with a positive reaction, as already reported in studies carried out by Owen, Willis and Lamph (2010), Johnson, Mills and Bezzole (2009) and Colón-Reveles et al. (2007).

Silbemagel and Lindberg (2003). These researchers obtained accurate results in a repeatability study on various food samples using Petrifilm ™ and conventional methodology, and recommend its use in counting enterobacteria for various foods, including frozen prepared meat. In this study, positive results were obtained with r= 0.83 agreement between the Petrifilm ™ EB and the conventional methodology for enterobacteria in *swabs* from pork carcasses, in agreement with the study by Silbemagel and Lindberg (2002), in which the Petrifilm ™ EB was more sensitive and selective compared to the sensitivity of the conventional method on a plate with Violet Red Bile Glucose (VRBG) agar. Both alternative methods used in this study were sensitive and specific in detecting enterobacteria.

There was a medium correlation between the conventional method and Tempo® (R= 0.67) for mesophilic microorganisms. These results represent a lower correlation compared to the evaluation study carried out by Green et al. (2007), when r=0.97 was found between the Tempo® method and the conventional plate method for counting mesophilic microorganisms. The same was observed in relation to the study carried out by Devulder et al. (2007) and Crowley et al. (2009) who found r=0.95 agreement for the same methodologies. In this study, a strong correlation was observed between the conventional methodology and Petrifilm™ (R= 0.77) and between the two alternative methodologies Tempo® and Petrifilm™ (R= 0.81).

4.3 CONCLUSIONS

Although some research using the alternative methods studied has already been published, the results of this study contributed to the evaluation of these methodologies in pork matrices that had not previously been studied.

In this study, the samples came from swabs of pork carcasses processed in a slaughterhouse under Federal Inspection with all the Good Manufacturing Practices and quality programmes in place, which gave the samples microbiological quality and low numbers of *Escherichia coli* detected in the methodologies evaluated.

The TEMPO® system methodology showed an advantage over the other methodologies evaluated in this study, because it processes a greater number of samples per time interval, reduces the volume of material and space needed to carry out the tests and because it is automated, which speeds up the process. However, the

Petrifilm ™ system methodology showed a slightly better response in terms of counting mesophilic microorganisms and enterobacteria.

Both methods can be recommended for the pork matrix, thus benefiting the processing industry which needs rapid responses in its production process.

CHAPTER 5

INVESTIGATION OF *ESCHERICHIA COLI* O157:H7 IN PIG CARCASSES IN THE STATE OF SANTA CATARINA, BRAZIL

Since it was identified in 1982 in the USA, *Escherichia coli* O157:H7 has become an important pathogen and the subject of much research. The cause of several outbreaks of food-borne infections associated with bloody diarrhoea (DOYLE et al., 2007), it is the most prevalent serotype in the Enterohaemorrhagic *E. coli* (EHEC) group. Its symptoms range from mild diarrhoea to haemorrhagic colitis (HC), haemolytic uremic syndrome (HUS) and thrombotic thrombocytopenic purpura (TTP) in humans (CONEDERA et al., 2007; O'LOUGHLIN; ROBINS-BROWNE, 2001). All EHEC produce cytotoxicity factors called verotoxin (VT) or shiga-like toxins (SLTs), which are toxic proteins that are biologically, structurally and antigenically similar to the shiga-toxin (ShT) produced by *Shigella dysenteriae* type 1 (AGBODAZE, 1999; LAW, 2000; MENG et al, 2007; MENG; ZHAO; DOYLE, 1998), associated with outbreaks in humans (KARMALI; GANNON; SARGEANT, 2010).

Studies show that the intestinal tract of cattle is the main reservoir of *E. coli* O157:H7, however this bacterium has been isolated in other ruminants such as sheep, goats and, sporadically, in chickens and pigs (AL-GALLAS et al., 2002; ATEBA; BEZUIDENHOUT, 2008; SILVA, A. et al., 2001). As an agent of food-borne illnesses, it has been particularly associated with the consumption of undercooked beef, unpasteurised milk and drinking and recreational water contaminated by the intestinal contents of animals (ACHESON et al., 1996; ATEBA; MBEWE, 2011; HUSSEIN; BOLLINGER, 2005; SANDRINI et al., 2007).

Escherichia coli O157:H7 has many characteristics typical of *E. coli,* but it has some specificities such as the absence or low fermentation of sorbitol and the absence of P-D-glucoronidase enzyme activity (MENG et al., 2007).

Rapid methods for detecting *Escherichia coli* O157 from food are commercially available, many of which are based on enzyme-linked immunosorbent assay and nucleic acid hybridisation technologies. The detection limits (typically 10^5 CFU/mL for ELISA) are reached after a pre-enrichment step that allows the target microorganisms to be recovered. Qualitative detection of *Escherichia coli* O157 can be carried out using the automated Vidas® system test and in human food products using the ELFA (Enzyme Linked Fluorescent Assay) technique (BLACKBURN; MCCARTHY, 2000).

Given that *Escherichia coli* O157:H7 has been naturally found in various foods such as fresh vegetables (HUANG et al., 2005), meat and dairy products (Al-GALLAS et al., 2002; HUSSEIN; BOLLINGER, 2005), milk and drinking water (CLOUGH; CLANCY; FRENCH, 2006; SANDRINI et al., 2007), beef carcasses (FEGAN et al., 2005), pork and pork products (ATEBA; MBEWE, 2011; OTEIZA et al., 2006), the aim of this study was to investigate *Escherichia coli* O157:H7 in pork carcasses produced in the state of Santa Catarina, Brazil.

4.4   SAMPLES

From March 2010 to March 2011, 152 samples were analysed from 76 pig carcasses collected at the end of the line at a Federally Inspected slaughterhouse, where 4,500 pigs were slaughtered every day. Each sample consisted of sponge *swabs* collected from four points on the carcass (shank, belly, loin and jowl) and packed in a single sterile *stomacher* bag. The four sponge *swabs* together (100 cm each) represent the sampling of an area corresponding to 400 cm^2 . Samples from the surfaces of the carcasses were collected by pressing the sponge, previously moistened in 100 mL of sterile 0.1% peptone saline solution, onto sterile delimiter moulds in the area to be sampled (100 cm) for around 20 seconds, starting the procedure vertically, then rubbing horizontally and finally diagonally. The sponge was placed in a sterile *stomacher* bag and kept refrigerated at 4 °C until it was analysed in the laboratory, which was no more than 24 hours after collection.

4.5   DETECTION OF *ESCHERICHIA COLI* O157:H7

The Vidas® system (Biomérieux S.A.) methodology was used to identify *Escherichia coli* O157:H7, and the conventional methodology was used as a subsequent confirmation step (AMERICAN PUBLIC HEALTH ASSOCIATION, 2001).The flowchart of the methodology used in the Vidas® system (Biomérieux S.A.) to detect *Escherichia coli* O157:H7 in pig carcasses is shown in Figure 12. Following the protocol recommended by the manufacturer, the positive samples were sent for confirmation using the conventional methodology, slide agglutination technique with *anti-Escherichia coli* 0157 serum (Probac do Brasil, SP) and PCR analysis carried out by the Enterobacteria Department of the Oswaldo Cruz Foundation, located in the city of Rio de Janeiro - Brazil.

The conventional methodology used was described by the *American Public Health Association* - APHA (2001), after which five to ten typical colonies were selected for biochemical and genotypic confirmation by PCR.

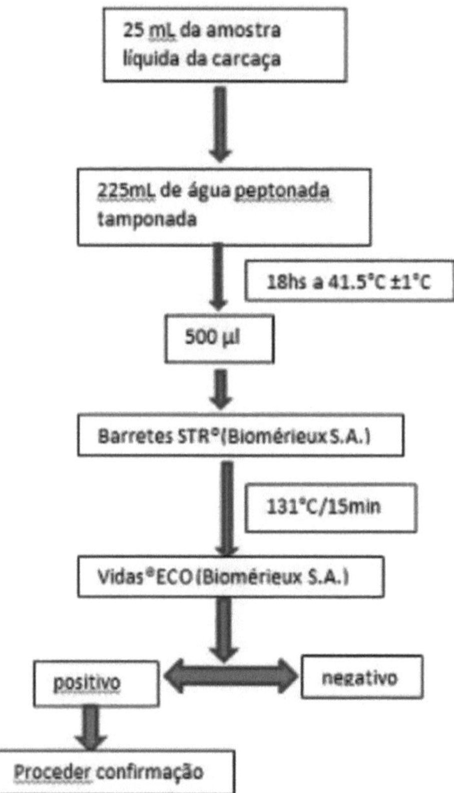

Figure 12 - Flowchart of the steps used in the methodology of the Vidas® system (Biomérieux S.A.) for detecting *Escherichia coli* O157:H7 in pig carcasses

Source: Biomerieux (2008)

5.   3RESULTS

For the pig carcass samples, three positive results were observed using the Vidas® system. These positive samples were plated on TC-SMAC and typical translucent colonies with a greyish centre (1-2 mm) and serum agglutination of the colonies were observed, but there was no confirmation by PCR and the virulence genes (r/&O157, *stx\, stx2)* were not found. Genotypic identification revealed *Salmonella* serotype Agona.

5.   4DISCUSSION AND CONCLUSIONS

In Brazil, there are few studies characterising *Escherichia coli* O157:H7 in animal products and their results have detected low incidence or absence of the pathogen (JAKABI et al, 2004; PRATA, 2009; SANDRINI et al, 2007; SILVA, A. et al., 2001; SILVEIRA, 2010). Shiga toxin-producing *Escherichia coli* strains have been found to be more prevalent in the faeces and serum of Brazilian cattle herds (CERQUEIRA et al., 1999; IRINO et al., 2005; LEOMIL et al., 2003).

The results of this research with samples of pork carcasses produced in the state of Santa Catarina reveal a low frequency of *Escherichia coli* O157:H7 in these products and are in line with other studies carried out in Brazil, in which the incidence of *Escherichia coli* was low or zero (JAKABI et al., 2004; PRATA, 2009; SANDRINI et al., 2007; SILVA, A. et al, 2001; SILVEIRA, 2010).

The lack of positive results for *Escherichia coli* O157:H7 in animal products in Brazil differs from the results found in food from various countries.

In New Zealand, Wong, MacDiarmid and Cook (2009) found a prevalence of 1% in 100 samples of pork carcasses. In South Africa, Ateba and Mbewe (2011) confirmed 130 isolates of *Escherichia coli* O157:H7 in 220 samples, the prevalence was highest in pork and pork products (67.7 per cent). In Spain, Mora et al. (2005) examined 722 STEC isolates, all recovered from humans, cattle, sheep and food during the years 1992-1999, 141 of which were identified as *Escherichia coli* O157:H7. In Ireland, Camey et al. (2006) analysed 1351 beef clippings, 131 carcasses and 132 meat from bovine heads and from these *Escherichia coli* O157:H7 was recovered from 2.4% (32/1351) of clippings

of beef, 3.0 per cent (4/132) of carcasses and 3.0 per cent (3/100) of head meat.

Vemozy-Rozand et al. (2002) carried out a study on 3,450 samples of minced beef, 175 of which tested positive using the Vidas *E.coli* 0157 (ELFA) method. Of these, 4 colonies were confirmed by Vidas ICE and identified as sorbitol negative and vero toxin producing, confirmed as O157-positive, H7- positive. In Mexico, Varela-Hemández et al. (2007) analysed 258 bovine carcass samples and 2.7% (7/258) were identified as *Escherichia coli* O157:H7.

In Australia, Fegan et al. (2005) analysed 606 samples. Of this total, 100 were carcass swabs with 6 % positive results for *Escherichia coli* O157:H7. In Tunisia, Al-Gallas et al. (2002) analysed 204 samples of meat and dairy products and isolated 3 *Escherichia coli* from the STEC group. In Nigeria, Ojo et al. (2010) found 7.3 per cent (154/2133) of *Escherichia coli* strains belonging to the STEC group, of which 4 per cent was detected in pork, 3.8 per cent in beef and 1.7 per cent in mutton. The detection rate of *Escherichia coli* O157:H7 was 5% in the 2,133 samples.

Lee et al. (2009), in Korea, analysed 3,000 samples of beef, pork and chicken and found 273 samples positive for *Escherichia coli,* 35.9% of which were classified as belonging to the EHEC group. In Austria, Halabi et al. (2008) investigated 2633 water samples, 280 were positive for *Escherichia coli,* and of these 3.9% (11/280) presented the virulence factors present in the EHEC group. In the USA, Hussein and Bollinger (2005) published a review of articles on contaminated beef. This review assessed the prevalence rates and public health risks of contamination by *Escherichia coli* belonging to the STEC group. Prevalence rates of 0157 ranging from 0.01 to 54.2 per cent were found.

In Turkey, Sarimehmetoglu et al. (2009) analysed 251 samples of fresh ground beef and 7.6% (19/251) were positive for *Escherichia coli* O157:H7. Çadirci et al. (2010) analysed 200 samples of fresh beef and detected *Escherichia coli* 0157 in 2.5% (5/200) of them.

In Italy, Conedera et al. (2007) analysed 3879 food samples for the presence of *Escherichia coli* 0157 (VETC), of which 931 were minced beef and 2948 dairy products (pasteurised milk, unpasteurised milk and cheese). *Escherichia coli* 0157 was isolated in 0.43 per cent (4/931) of minced beef and one 0157 strain in cheese. This strain did not have the *stxl* and *s1x2* genes, but the *eae* gene, indicating pathogenicity. In Argentina, Oteiza et al. (2006) analysed 100 pre-cooked sausages (morcilla), in 2% of the samples *Escherichia coli* O157:H7 was identified.

In Greece, Dontorou et al. (2003) carried out a survey of 600 samples of unpasteurised milk, raw minced meat, beef hamburger, sandwiches, salad, traditional Greek sausage and prepared pork intestines with the aim of detecting the presence of *Escherichia coli* O157:H7. The pathogen was detected in 1% (1/100) of milk samples and 1.3% (1/75) of fresh sausages and 2% (1/50) of prepared pork intestines. Chye, Abdullah and Ayob (2004) also carried out a study with 930 samples of raw milk collected in tanks from 360 dairy farms in Malaysia, *Escherichia coli* O157:H7 was detected in 33.5 per cent (312/930) of the 930 samples.

Huang et al. (2005) carried out research on fresh cut vegetables artificially inoculated with *Escherichia coli* 0157 and analysed using the Vidas® system methodology, and the results showed that 40% of the control samples were positive for *Escherichia coli* 0157 in the Vidas® system, while no positive results were found by multiplex PCR, thus concluding that the false-positive results found in the Vidas® system were affected by *Morganela morganii and Enterobacter cloacae* due to antigenic similarity.

A similar result was found in this study in the pork carcass samples, in which three results were positive in the Vidas® system, but this was a false positive result. This is because some *Salmonella* serotypes can interfere with the methodology of the ELFA technique, generating false-positive results, since they have an antigenic capsule (O antigen) similar to that found in *Escherichia coli* 0157 (AMERICAN PUBLIC HEALTH ASSOCIATION, 2001; HUANG et al., 2005).

The negative results for *Escherichia coli* O157:H7 found in this study are conclusive in indicating the absence of the pathogen in the pig carcasses sampled in the state of Santa Catarina. Data from this study cannot be compared with data from other studies carried out in Brazil due to the variability of samples and methodological techniques used in the tests. So far, it has not been possible to establish the incidence and prevalence of *Escherichia coli* O157:H7 in animal products in Brazil, unlike in other countries.

CHAPTER 6

FINAL CONSIDERATIONS

The first part of this study consisted of the application of decontamination methods and was carried out in an establishment considered to be a model for pig slaughter in the state of Santa Catarina, Brazil. The industrial design of this slaughterhouse follows a straight-line flow, which allows us to avoid counter-flow during the slaughter process and thus minimise cross-contamination between carcasses. Checks on the application of Good Manufacturing Practices and operational hygiene control programmes demonstrated the establishment's good level of sanitation, which was confirmed by the results of the analyses with low levels of contamination by mesophilic microorganisms, enterobacteria and *E. coli* in the carcasses sampled.

The occurrence of *Salmonella* spp. in the carcasses sampled demonstrates the need to know the prevalence of this pathogen in the pig herd when it enters slaughter, so that parameters for the maximum limit of the pathogen in pig carcasses can be set by Ministry of Agriculture legislation, and thus qualify and quantify the pathogen and its real significance in terms of health risk.

The absence of positive results for *E. coli* 0157 in the pig carcasses sampled showed that we may be facing low incidences of the pathogen in the pig herd, but there is a need for continuous monitoring so that, if detected, the danger can be controlled in time to avoid jeopardising public health.

The focus when using decontamination procedures is on reducing pathogenic microorganisms. In this study, due to the scarcity of positive results for pathogenic micro-organisms after the application of all the decontamination treatments used, there was insufficient support for a more detailed evaluation. The scientific literature mentions the need to use decontamination procedures on carcasses with high levels of microbial contamination at the end of slaughter, which could not be verified in this study.

There is a health gain in the post-slaughter stages by reducing the microbial load of the carcasses.

It can be concluded that the prevalence of indicator and pathogenic microorganisms that may be present in pig carcasses during the slaughter process is dependent not only on the microbial load at the entrance to the slaughterhouse, but also on the procedures adopted during slaughter.

It is understood that in certain circumstances, when the pathogenic microbial load entering the slaughterhouse is known, there is a need to introduce decontamination methods in pig carcasses, with the aim of reducing pathogens.

The second part of this study consisted of microbiological analyses using different methodologies and was carried out at the food microbiology laboratory of the Federal University of Santa Catarina, which has all the necessary requirements for carrying out the tests and is accredited by the Ministry of Agriculture, Livestock and Supply.

The results of the conventional methodologies used were compared with the results of the alternative TEMPO® and PETRIFILM™ methodologies, and positive correlations were obtained in the different tests carried out.

The TEMPO® system processes 500 tests in 4 hours and results for mesophilic microorganisms, enterobacteria and *E. coli* are obtained between 24 and 48 hours. The PETRIFILM™ system reduces steps compared to the conventional system and results are also obtained between 24 and 48 hours.

The use of alternative methodologies should be encouraged and utilised, as there are significant savings in material, labour and a reduction in the time it takes to carry out the tests until the results are obtained, which means a gain for the laboratories that carry out the tests, for the industry that needs fast and reliable results, and for the population that receives safe products of animal origin for consumption.

REFERENCES

ACHA, P. N.; SZYFRES, B. **Zoonoses and communicable diseases common to man and animals.** 3. ed. Washington, DC: Pan American Health Organisation, 2003. v. 1.

ACHESON, D. W. K. et al. Detection of Shiga-like toxin-producing *Escherichia coli* in ground beef and milk by commercial enzyme immunoassay. **Journal of Food Protection,** Des Moines, IA, v. 59, n. 4, p. 344-349, 1996.

AFNOR VALIDATION. **Validated methods.** Available at: <www.afiior-validation.com>. Accessed on: 20 May 2011.

AGBODAZE, D. Verocytotoxins (Shiga-like toxins) produced by *Escherichia coli:* a minireview of their classification, clinical presentations and management of a heterogeneous family of cytotoxins. **Comparative Immunology, Microbiology & Infectious Diseases,** Amsterdam, The Netherlands, v. 22, n. 4, p. 221-230, 1999.

AL-GALLAS, N. et al. A. Isolation and characterisation of Shiga toxin- producing *Escherichia coli* from meat and dairy products. **International Journal of Food Microbiology,** Amsterdam, The Netherlands, v. 19, p. 389-398, 2002.

ALMEIDA, I. A. Z. C. et al. *Salmonella:* serotypes identified in the region of São José do Rio Preto/SP, in the period 1990-1999. **Revista do Instituto Adolfo Lutz,** São Paulo, n. 59, p. 33-37, 2000.

AMERICAN PUBLIC HEALTH ASSOCIATION (APHA).
Compendium of Methods for the Microbiological Examination of Foods. 4. ed. Washington, DC, 2001.

ANDRADE, D. F.; OGLIARI, P. J. **Statistics for agricultural and biological sciences.** Florianópolis: UFSC, 2007.

ANDREWS, W. H. et al. *Salmonella.* In: AMERICAN PUBLIC HEALTH

ASSOCIATION (APHA). **Compendium of Methods for the Microbiological Examination of Foods.** 4. ed. Washington, DC, 2001. p. 357-380.

ANIL, M. H.; WHITTINGTON, P. E.; MCKINSTRY J. L. The efect of the sticking method on the welfare of slaughter pigs. **Meat Science,** Amsterdam, Netherlands, n. 55, p. 315-319, 2000.

BRAZILIAN PORK PRODUCING AND EXPORTING INDUSTRY ASSOCIATION. **Brazilian pork.** São Paulo, 2011. Available at: <http://www.abipecs.org.br/uploads/relatorios/relatorios- associados/ABIPECS relatorio 2010 pt.pdf>. Accessed on: 10 January 2012.

OFFICIAL ANALYTICAL CHEMISTS ASSOCIATION (OAC). **Official Methods of Analysis.** 17. ed. Washington, DC, 2002. v. 1.

ASSOCIATION OFFICIAL ANALYTICAL CHEMISTS RESERCH NEWS. TEMPO® EB Test Granted PTM Status. **Inside Laboratory Management,** Marcy l'Etoile, France, may/june 2009.

ATEBA, C. N.; BEZUIDENHOUT, C. C. Characterisation of *Escherichia coli* 0157 strains from humans, cattle and pigs in the Northwest Province, South Africa. **International Journal of Food Microbiology,** Amsterdam, The Netherlands, v. 128, p. 181-188, 2008.

ATEBA, C. N.; MBEWE, M. Detection of *Escherichia coli* O157:H7 virulence genes in isolates from beef, pork, water, human and animal species in the northwest province, South Africa: public health implications. **Research in Microbiology,** Amsterdam, The Netherlands, p. 1-9, 2011.

AYMERICH, T.; PICOUET, P. A.; MONFORT, J. M. Decontamination technologies for meat products. **Meat Science,** Amsterdam, The Netherlands, n. 78, p. 114-129, 2008.

BAGGESEN, D. L. et al. Herd prevalence of *Salmonella* enterica infections in Danish slaughter pigs determined by microbiological testing. **Preventive veterinary medicine,** Amsterdam, The Netherlands, n. 26, p. 201-213, 1996.

BEARSON, B. L.; BEARSON, S. M. D. Host specific differences alter the requirement for certain *Salmonella* genes during swine colonisation. **Veterinary Microbiology,** Amsterdam, The Netherlands, v. 150, n. 3-4, p. 215- 219, jun. 2011.

BELK, K. E. Beef decontamination technologies. **Beef Facts,** Centeimial, CO, v. 11, p. 1-4, 2002.

BERENDS, B. R. et al. Identification and quantification of rosk factors regarding *Salmonella* spp. on pork carcasses. **International Journal of Food Microbiology,** Amsterdam, The Netherlands, v. 36, p. 199-206, 1997.

BERGEY, D. **Bergey's manual of determinative bacteriology.** 7. ed. Baltimore: Williams & Wilkins, 1994.

BESSA, M. C.; COSTA, M.; CARDOSO, M. Prevalence of *Salmonella spp* in pigs slaughtered in slaughterhouses in Rio Grande do Sul. **Pesquisa Veterinária Brasileira,** Seropédica, v. 24, n. 2, p. 80-84, Apr./Jun. 2004.

BINTER, C. et al. Transmission and control of *Salmonella* in the pig feed chain: a conceptual model. **International Journal of Food Microbiology,** Amsterdam, The Netherlands, v. 145, p. 7-17, Mar. 2011.

BIOMERIEUX. **Training manual.** Paris, France, 2008.

BLACKBURN, C. W.; MCCARTHY, J. D. Modifications to methods for the enumeration and detection of injured *Escherichia coli* O157:H7 in foods. **International Journal of Food Microbiology,** Amsterdam, The Netherlands, v. 55, p. 285-290, 2000.

BLAHA, T. G. Pre-harvest food safety and *Salmonella* reduction in the pork chain. In: ABRAVES CONGRESS, 10, 2001, Porto Alegre. **Proceedings...** Porto Alegre, 2001.

BLOCH, N. et al. The enumeration of Coliforms and E. Coli on naturally contaminated beef: a comparison of the Petrifilm™ Method with the Australian Standard. **Meat Science,** Amsterdam, The Netherlands, v. 43, n. 2, p. 187-193, 1996.

BLOOD, R. M.; CURTIS, G. D. W. Media for total *Enterobacteriaceae,* Coliforms and *Escherichia coli.* **International Journal of Food Microbiology,** Amsterdam, The Netherlands, v. 26, p. 93-115, 1995.

BOER, E. D.; BEUMER, R. R. Methodology for detection and typing of foodbome microorganisms. **International Journal of Food**

Microbiology, Amsterdam, The Netherlands, v. 50, n. 1-2, p. 119-130, 1999. Available at: <http://www.sciencedirect.coni/science/article/pii/S0168160599000811>. Accessed on: 10 June 2011.

BOLTON, D. J.; DOHERTY, A. M.; SHERIDAN, J. J. Beef HACCP: intervention and non-intervention systems. **International Journal of Food Microbiology,** Amsterdam, The Netherlands, v. 66, p. 119-129, 2001.

BOLTON, D. J.; PEARCE, R.; SHERIDAN, J. Risk based determination of critical control points for pork slaughter. **Food Safety,** Ashtown, Dublin, 2002.

BOLTON, D. J. et al. Decontamination of pork carcasses during scalding and the prevention of *Salmonella* cross-contamination. **Journal of Applied Microbiology,** New Jersey, n. 94, p. 1036-1042, 2003.

BOSILEVAC, J. M. et al. Treatments using hot water instead of lactic acid reduce levels of aerobic bacteria and *Enterobacteriaceae* and reduce the prevalence of *Escherichia coli* O157:H7 on preevisceration beef carcasses. **Journal of Food Protection,** Des Moines, IA, v. 69, n. 8, p. 1808-1813,2006.

BRAZIL. Ministry of Agriculture and Agrarian Reform. Decree no. 30691, of 29 March 1952. Approves the new Regulations for the industrial and sanitary inspection of products of animal origin. **Diário Oficial da União,** Brasília, DF, 07 July 1952. Section I, p. 10785.

BRAZIL. Ministry of Agriculture and Supply. Ordinance no. 368, of 4th September 1997. Approves the Technical Regulation on Hygienic-Sanitary Conditions and Good Manufacturing Practices for Food Processing/Industrialising Establishments. **Diário Oficial da União,** Brasília, DF, 08 Sep. 1997. Section I, p. 19697.

BRAZIL. Ministry of Agriculture, Livestock and Supply. **Circular no. 130 of 13 February 2007.** Brasília, DF, 2007.
Pork exports to the Member States of the European Union. Available at: <http://sigsif.agricultura.gov.br/sigsif7principal sigsif>. Accessed on: 11 June 2011.

BRAZIL. Ministry of Agriculture, Livestock and Supply. **Circular no. 640/2004/DCI/DIPOA of 22 October 2004.**
Brasilia, DF, 2004. Use of carcass decontaminant in establishments authorised to export.

BRAZIL. Ministry of Agriculture, Livestock and Supply. Normative Instruction No. 40 of 12 December 2005. Approves the Analytical Methods, Isolation and Identification of *Salmonella* in beef, poultry and egg products - MLG - 4.03, Alternative Methodology for *Salmonella* A-Bax -MLG 4C .01, Isolation and Identification of Listeria Monocytogenes in red meat, poultry meat, eggs and environmental samples, MLG 8.04 - Alternative Methodology for Listeria A-BAX MLG-8 A .01, *Escherichia coli*, MPN AO AC 966.24, Petrifilm Method AO AC 998.08, which become Official Standards for Analysing the Microbiology of Products of Animal Origin. **Diário Oficial da União,** Brasília, DF, 16 Dec. 2005. Section I, p. 70. Available at: <http://extranet.agricultura.gov.br/sislegis/action/detalhaAto.do7method =consultarLegislacaoFederal>. Accessed on: 29 Nov. 2011.

BRAZIL. Ministry of Agriculture, Livestock and Supply. Normative Instruction No. 62 of 26 August 2003. Official Analytical Methods for Microbiological Analyses for the control of products of animal origin and water. **Federal Official Gazette,** Brasília, DF, 18 September 2003. Section 1, p. 14.

BRAZIL. Ministry of Agriculture, Livestock and Supply. Ordinance no. 46, of 10 February 1998. Establishes the system for analysing hazards and critical control points: HACCP, to be implemented in industries producing products of animal origin. **Diário Oficial da União,** Brasília, DF, 10 Feb. 1998. Section I.

BRAZIL. Ministry of Agriculture, Livestock and Supply. Ordinance No. 711, of 1st November 1995. Approves the technical standards for facilities and equipment for slaughtering and industrialising pigs.
Diário Oficial da União, Brasília, DF, 03 Nov. 1995. Section I, p. 17625.

BRAZIL. Ministry of Health. **Epidemiological analysis of food-borne disease outbreaks in Brazil.** Brasília, DF, 2010. Available at:

<http://portal.saude.gov.br/portal/saude/profissional/visualizar texto.cf m?idtxt=31758>. Accessed on: 18 August 2011.

BRAZIL. Ministry of Health. Ordinance No. 2914 of 12 December 2011. Provides for procedures to control and monitor the quality of water for human consumption and its standard of potability. **Federal Official Gazette,** Brasília, DF, 04 January 2012. Section I, p. 43-49.

ÇADIRCI, Õ. et al. The prevalence of *Escherichia coli* 0157 and 0157: H7 in ground beef and raw meatball by immunomagnetic separation and the detection of virulence genes using multiplex PCR. **Meat Science,** Amsterdam, The Netherlands, v. 84, p. 553-556, 2010.

CAREY, C. M. et al. The effect of probiotics and organic acids on Shiga-toxin 2 gene expression in enterohemorrhagic *Escherichia coli* O157:H7. **Journal of Microbiological Methods,** Amsterdam, The Netherlands, n. 73, p. 125-132, 2008.

CARNEY, E. O'BRIEN et al. Prevalence and levei of *Escherichia coli* 0157 on beef trimmings, carcasses and boned head meat at a beef slaughter plant. **International Journal of Food Microbiology,** Amsterdam, The Netherlands, v. 23, p. 52-59, 2006.

CARPENTER, C. E.; SMITH, J. V.; BROADBENT, J. R. Efficacy of washing meat surfaces with 2% levulinic, acetic, or lactic acid for pathogen decontamination and residual growth inhibition. **Meat Science,** Amsterdam, The Netherlands, n. 88, p. 256-260, 2011.

CASAROTTI, S. N.; PAULA, A. T.; ROSSI, D. A. Correlation between chromogenic methods and the conventional method in the enumeration of coliforms and *Escherichia coli* in ground beef. **Revista do Instituto Adolfo Lutz,** São Paulo, v. 3, n. 66, p. 278-286, 2007.

CATTANI, C. S. de O. **Photos from personal archive.** Chapecó, 2010.

CATTANI, C. S. de O. **Photos from personal archive.** Florianópolis, 2011.

CATTANI, C. S. de O. **Occurrence of *Salmonella* spp. and *Listeria monocytogenes* isolated in pork products in the state of Santa Catarina, from 2007 to 2010.** Dissertation

Professional in Hygiene, Inspection and Technology of Food of Animal Origin)- Faculty of Veterinary Science, Fluminense Federal University, Niterói, 2011.

CATTAPAN, F.; VILLARD, P.; DUMONT, N. **A new approach for workflow evaluation in food microbiology laboratories: TEMPO,** Automated MPN Method Versus Plate Count Method. Paris, France: bioMérieux, 2009. Available at: <http ://www.biomerieux- usa.com/upload/TEMPQ%20workflow%20CHELAB-1 .pdf>. Accessed on: 12 May 2011.

CATTAPAN, F. et al. **New approach for workflow evaluation in food microbiology laboratories:** TEMPO®, Automated MPN Method Versus Plate Count Method. Paris, France: bioMérieux, 2007. Available at: <http://www.foodprotection.org/files/annual meeting/poster-abstracts- 2007.pdf>. Accessed on: 05 May 2011.

CERQUEIRA, A. M. F. et al. High occurrence of Shiga toxin-producing *Escherichia coli* (STEC) in healthy cattle in Rio de Janeiro State, Brazil. **Veterinary Microbiology,** Amsterdam, The Netherlands, v. 70, p.111-121,1999.

CHYE, F. Y.; ABDULLAH, A.; AYOB, M. K. Bacteriological quality and safety of raw milk in Malaysia. **International Journal of Food Microbiology,** Amsterdam, The Netherlands, v. 21, p. 535-541, 2004.

CLOUGH, H. E.; CLANCY, D.; FRENCH, N. P. Vero-Cytotoxigenic *Escherichia coli* 0157 in pasteurised milk containers at the point of retail: a qualitative approach to exposure assessment. **Risk Analysis,** New Jersey, v. 26, n.5, p. 1291-1309, 2006.

CODE of Hygienic Practice for Meat: CAC/RCP 58-2005. 2009. Available at: <www.codexalimentarius.net>. Accessed on: 11 July 2011.

CODE of Hygienic Practice for Meat: draft code of hygienic practice for meat. FAO, 2004a. ALINORM 04/27/16.

CODE of Hygienic Practice for Meat: regulation (EC) n° 853/2004 of the European Parliament and the Concil of 29 april 2004, laying down

specific hygiene rules for food of animal origin. **Official Journal of the European Union,** European Union, 30 apr. 2004b.

COLÓN-REVELES, J. et al. **Evaluation of the TEMPO® EB Method for the Enumeration of Enterobacteriaceae in foods.** Durham: bioMérieux, 2007. p. 3-47. Available at: <http://www.biomerieuxusa.com/upload/TEMPQ%20EB%20bMx%20 US-2.pdf> . Accessed on: 02 May 2011.

CONEDERA, G. et al. A Family outbreak of *Escherichia coli* 0157 haemorrhagic colitis caused by pork meat salami. **Epidemiology Infect,** Cambridge, UK, v. 135, p. 311-314, 2007.

CONTROL of hazards to human and animal health through ante-mortem and post-mortem meat inspection. Paris, France, 2006. Information document drawn up by the OIE Working Group on the Health Safety of Food Derived from Animal Production.

CORNICK, N. A.; HELGERSON, A. F. Transmission and infectious dose of *Escherichia coli* O157:H7 in swine. **Applied and Environmental Microbiology,** Washington, DC, v. 70, n. 9, p. 5331- 5335, sept. 2004.

CORTINAS A. J. et al. *Salmonella serosurvaillance* different statistical methods to categorise pig herds based on serological data. **Preventive Veterinary Medicine,** Amsterdam, The Netherlands, n. 89, p. 59-66, 2009.

CROWLEY, E. S. et al. TEMPO®EC for the enumeration of *Escherichia coli* in foods: collaborative study. **Journal of AOAC International,** Gaithersburg, MD, v. 93, p. 576-586, 2010. Available at: <http://www.ncbi.nlm.nih.gov/pubmed>. Accessed on: 05 May 2011.

CROWLEY, E. S. et al. TEMPO®TVC for the enumeration of aerobic mesophilic flora in foods: collaborative study. **Journal of AOAC International,** Gaithersburg, MD, v. 92, n. 1, p. 165-174, 2009. Available at: <http://www.foodhygiene.or.kr/admin/down file.php.pdf>. Accessed on: 05 May 2011.

D'AOUST, J.; MAURER, J.; BAILEY, J. S. *Salmonella* Species. In: DOYLE, M. P.; BEUCHAT, L. R. **Food Microbiology:** fundamentals and frontiers. 3. ed. Washington, DC: Montville, 2007.

DALLACOSTA, O. A. et al. Evaluation of transport and unloading conditions and the occurrence of falls in pigs from an animal welfare perspective. **Technical communiqué,** Concórdia, n. 459, Dec. 2007.

DAVIDSON, P. M.; TAYLOR T. M. Chemical preservatives and natural antimicrobial compounds. In: DOYLE, M. P.; BEUCHAT, L. R. **Food microbiology:** fundamentals and frontiers. 3. ed. Washington, D.C.: Montville, 2007.

DE BOER, E.; HEUVELINK, A. E. Methods for the detection and isolation of Shiga toxin-producing *Escherichia coli.* **Journal of Applied Microbiology,** New Jersey, v. 88, p. 133-143, 2000

DE BUSSER, E. V. et al. Detection and characterisation of *Salmonella* in lairage, on pig carcasses and intestines in five slaughterhouses. **International Journal of Food Microbiology,** Amsterdam, The Netherlands, v. 145, n. l,p. 275-286, 2011.

DEVULDER, G. et al. **Comparison of TEMPO® TVC, TEMPO® TC, TEMPO® EC with conventional culture methods in Meat Based Products.** Paris, France: bioMérieux, 2007. Available at: <http://www.biomerieux-usa.com/upload/TEMPO.pdf>. Accessed on: 02 May 2011.

DONTOROU, C. et al. Isolation of *Escherichia coli* O157:H7 from foods in Greece. **International Journal of Food Microbiology,** Amsterdam, The Netherlands, v. 82, p. 273-279, 2003.

DORSA, W. J. et al. Microbial decontamination of beef and sheep carcasses by steam, hot water spray washes, and a steam-vacuum sanitizer. **Journal of food protection,** Des Moines, IA, v. 59, n. 2, p. 127-135, 1995.

DOYLE, M. et al. *E. coli* O157:H7 in food. In: DOYLE, M. P.; BEUCHAT, L. R. **Food Microbiology:** fundamentals and frontiers. 3. ed. Washington, D.C.: Montville, 2007. p. 171-192.

EPLING, L. K.; CARPENTER, J. A.; BLANKENSHIP, L. C. Prevalence of *Campylobacter* spp. and *Salmonella* spp. on pork carcasses and the reduction effected by spraying with lactic acid. **Journal of Food protection,** Des Moines, IA, v. 56, n. 6, p. 536-540, 1993.

UNITED STATES OF AMERICA. **Centers for disease control and prevention.** 2008. Available at: <http://www.cdc.gov/>. Accessed on: 3 May 2011.

UNITED STATES OF AMERICA. Department of Agriculture Food Safety and Inspection Service. **FSIS directive 7120.1:** revision 6: safe and suitable ingredients used in the production of meat, poultry, and egg products. Washington, DC, 2011.

UNITED STATES OF AMERICA. **Directive 7120.1 review from 4/6/2012:** safe and suitable ingredients used in the production of meat, poultry and eggs products. Available at: <http://www.fsis.usda.gov/OPPDE/rdad/FSISDirectives/712Q. 1 .pdf>. Accessed on: 10 August 2012.

UNITED STATES OF AMERICA. **Guidance on the procedures for joint Food Safety and Inspection Service and Food and Drug Administration:** approval of ingredients used in the production of meat and poultry products. 2000. Available at: <www.fsis.usda.gov/opped/larc>. Accessed on: 12 May 2011.

EUROPEAN COMMISSION. **Opinion of the scientific committee on veterinary mcasurcs rclating to public health on revision of meat inspection procedures.** European Union, 2000.

FAIRBROTHER, J. M. et al. Neonatal *Escherichia coli* diarrhoea. In: STRAW, B. E. et al. **Diseases of swine.** 8. ed. Iowa: AMES, 1999. p. 433-441

FEDORKA-CRAY, P. J.; GRAY, J. T.; WRAY, C. *Salmonella* infections in pigs. In: WRAY, C.; WRAY, A. (Ed.). ***Salmonella* in domestic animals.** Oxford: Oxford University, 2000. p. 191-207

FEGAN, N. et al. An investigation of *Escherichia coli* 0157 contamination of cattle during slaughter at an abattoir. **Journal of Food Protection,** Des Moines, LA, v. 68, n. 3, p. 451-457, 2005.

FELDSINE, P.; ABEYTA, C.; ANDREWS, W. H. AOAC International methods committee guidelines for validation of qualitative and quantitative food microbiological official methods of analysis. **Journal of AOAC International,** Gaithersburg, MD, v. 85, n. 5, p. 1187-1200, may 2002.

FERREIRA, M. F. et al. Effect of water spraying on carcass decontamination. In: INTERNATIONAL CONGRESS OF MEAT SCIENCE AND TECHNOLOGIES, 50, 2004, Helsinki, Finland. **Proceedings...** Helsinki, Finland, 2004.

FORSYTHE, S. J. **Microbiology of food safety.** Porto Alegre: Artmed, 2002.

FRANCO, B. D. G. M.; LANDGRAF, M. **Microbiologia dos alimentos.** São Paulo: Atheneu, 2005.

FRANSEN, N. G. et al. Pathogenic micro-organisms in slhaughterhouse sludge-a survey. **International Journal of Food Microbiology,** Amsterdam, The Netherlands, v. 33, n. 2-3, p. 245-256, 1996.

FREDRIKSSON-AHOMAA, M. et al. Prevalence of pathogenic Yersinia enterocolitica and Yersinia pseudotuberculosis in wild boars in Switzerland. **International Journal of Food Microbiology,** Amsterdam, The Netherlands, v. 135,

p. 199-202, 2009.

FRONDREVEZ, M. et al. Simplified method for detecting pathogenic *Yersinia enterocolitica* in slaughtered pig tonsils. **Journal of Microbiological Methods,** Amsterdam, The Netherlands, v. 83, n. 2, p. 244-249, sept. 2010.

FRYDENDAHL, K. Prevalence of serogroups and virulence genes in *Escherichia coli* associated with postweaning with diarrhoea and oedema disease in pigs and a comparison of diagnostic approaches. **Veterinary Microbiology,** Amsterdam, The Netherlands, n. 85, p. 169-182, 2002.

FUNG, D. Y. C. Rapid methods and automation in microbiology: 25 years of developments and predictions. **Alimentaria:** Revista de Investigación, Tecnologia y Seguridad, Madrid, p. 113-117, 2008.

GE, B.; MENG, J. Advanced technologies for pathogen and toxin detection in foods: current applications and future directions.
Technology review; JALA, St. Charles, IL, p. 235-241, aug. 2009.

GILL C. O.; BRYANT, J. Decontamination of carcasses by vacumm- hot water cleaning and steam pasteurising during routine operations at a beef packing plant. **Meat Science,** Amsterdam, The Netherlands, v. 47, n. 3-4, p. 267-276, 1997.

GILL, C. O.; LANDERS, C. Microbiological effects of carcass decontaminating treatments at four beef packing plants. **Meat Science,** Amsterdam, The Netherlands, n. 65, p. 1005-1011, 2003.

GILL, C.O. et al. Decontamination of commercial, polished pig carcasses with hot water. **International Journal of Food Microbiology,** Amsterdam, The Netherlands, v. 12, p. 143-149, 1995.

GREEN, R. et al. **Evaluation of the TEMPO® System for total viable count, total coliforms and *E. coli* Enumeration in Meat and Poultry Products.** Durham: bioMérieux, 2007. Available at:
<http://www.biomerieuxusa.com/upload/TVC%20TC%20and%20EC%20vs%20ISO%20in%20meat%20and%20poultry%20Campden-1 .pdf>. Accessed on: 12 May 2011.

GREGORY, N. G. Animal welfare at markets and during transport and slaughter. **Meat Science,** Amsterdam, Netherlands, n. 80, p. 2-11, 2008.

GUENTER, S. et al. Enterobacteriaceae populations during experimental *Salmonella* infection in pigs. **Veterinary Microbiology,** Amsterdam, The Netherlands, n. 142, p. 352-360, 2010.

HALABI, M. et al. Prevalence of Shiga toxin, intimin and haemolysin genes in *Escherichia coli* isolates from drinking water supplies in a rural area of Austria. **International Journal of Hygiene and Environmental Health,** Amsterdam, The Netherlands, v. 211, p. 454-457, 2008.

HARDIN, M. D. et al. Comparison of methods for decontamination from beef carcass surfaces. **Journal of Food protection,** Des Moines, IA, v. 58, p. 368-374, 1995.

HERRERA-LEÓN, S. et al. Molecular characterisation of a new serovar of *Salmonella bongori* 13,22:z39: isolated from a lizard. **Research in Microbiology,** Amsterdam, The Netherlands, n. 156, p. 597-602, 2005.

HEUVELINK, A. E. et al. Isolation and characterisation of verocytotoxin-producing *Escherichia coli* 0157 from slaughter pigs and poultry. **International Journal of Food Microbiology,** Amsterdam, The Netherlands, v. 52, n. 1-2, p. 67-75, 1999.

HUANG, C. et al. Development of a Modified Enrichment method for the Rapid Immunoassay of *Escherichia coli* 0157 strains in fresh cut vegetables. **Journal of Food and Drug Analysis,** Taiwan, v. 13. n. 1, p. 64-70, 2005.

HUFFMAN, R. D. Current and future technologies for the decontamination of carcasses and fresh meat. **Meat Science,** Amsterdam, Netherlands, n. 62, p. 285-294, 2002.

HUGAS, M.; TSIGARIDA, E. Pros and cons of carcass decontamination: the role of the European Food Safety Authority. **Meat Science,** Amsterdam, Netherlands, n. 78, p. 43-52, 2008.

HUIS int VELD, J. H. J.; MULDER, R. W. A. W.; SNIJDERS, J. M. A. Impact of animal husbandry and slaughter technologies on microbial contamination of meat: monitoring and control. **Meat Science,** Amsterdam, The Netherlands, v. 36, n. 1-2, p. 123-154, 1994.

HUMPHREY, T.; JORGENSEN, F. Pathogens on meat and infection in animals: establishing a relationship using *campylobacter* and *Salmonella* as examples. **Meat Science,** Amsterdam, Netherlands, n. 74, p. 89-97, 2006.

HUSSEIN, H. S.; BOLLINGER, L. M. Prevalence of Shiga toxin- producing *Escherichia coli* in beef. **Meat Science,** Amsterdam, The Netherlands, v. 71, p. 676-689, 2005.

HWANG, C.; BEAUCHAT, L. R. Efficacy of lactic acid/sodium benzoate wash solution in reducing bacterial contamination of raw food.

chicken. **International Journal of Food Microbiology,** Amsterdam, Netherlands, n. 27, p. 91-98, 1995.

INTERNATIONAL COMMISSION ON MICROBIOLOGICAL SPECIFICATIONS FOR FOODS. **Microorganisms in food 6:** Microbial ecology of food commodities. 2. ed. New York, 2005.

IRINO, K. et al. Serotypes and virulence markers of Shiga toxin- producing *Escherichia coli* (STEC) isolated from dairy cattle in São Paulo State, Brazil. **Veterinary Microbiology,** Amsterdam, The Netherlands, v. 105, n. l,p. 29-36, 2005.

JAKABI, M. et al. Occurrence of *Salmonella* spp. and *Escherichia coli* O157:H7 in raw meat marketed in São Paulo city, Brazil, and evaluation of its cold tolerance in ground beef. **Revista do Instituto Adolfo Lutz,** São Paulo, v. 63, n. 2, p. 238-242, 2004.

JASSON, V. et al. Alternative microbial methods: an overview and selection criteria. **International Journal of Food Microbiology,** Amsterdam, The Netherlands, v. 27, n. 6, p. 710-730, 2010.

JAYAWARDANA, B. C. et al. Removing of central nervous tissues from dressed carcasses: washing with a low concentration of lactic acid in spraying cabinet. **Food Control,** Amsterdam, The Netherlands, n. 20, p. 386- 390, 2009.

JENSEN, T.; CHRISTENSEN, H. Decontamination of pig carcasses with Hot Water. In: INTERNATIONAL CONGRESS OF MEAT SCIENCE AND TECHNOLOGIES, 50, 2004, Helsinki, Finland. **Proceedings...** Helsinki, Finland, 2004.

JOHNSON, R. BioMérieux TEMPO® EC Test Granted PTM Status. **Inside Laboratory Management,** Marcy l'Etoile, France, may/june 2007.

JOHNSON, R.; MILLS, J.; BEZZOLE, L. BioMérieux TEMPO® EB test Granted PTM Status. **Inside Laboratory Management,** Marcy l'Etoile, France, may/june 2009.

JOHNSON, R.; MILLS, J.; BEZZOLE, L. BioMérieux TEMPO® TVC test Granted PTM Status. **Inside Laboratory Management.** Marcy l'Etoile, France, may/june 2008.

JONES, T. C.; HUNT, R. D.; KING N. W. Digestive system. In: . **Veterinary pathology.** 6. ed. São Paulo: Manole, 2000. p. 1063-1130.

JORDAN, E. et al. *Salmonella* surveillance in raw and cooked meat and meat products in the Republic of Ireland from 2002 to 2004. **International Journal of Food Microbiology,** Amsterdam, Netherlands, n. 112, p.

66-70, 2006.

KARMALI, M. A.; GANNON, V.; SARGEANT, J. M. Verocytotoxin- producing *Escherichia coli* (VETC). **Veterinary Microbiology,** Amsterdam, The Netherlands, v. 140, p. 360-370, 2010.

KICH, J. D. et al. Factors associated with *Salmonella* seroprevalence in commercial pig herds. **Ciência Rural,** Santa Maria, v. 35, n. 2, p. 398-405, 2005.

KICH, J. D. et al. Prevalence, distribution and molecular characterisation of *Salmonella* recovered from swine fmishing herds and a slaughter facility in Santa Catarina, Brazil. **International Journal of Food Microbiology,** Amsterdam, The Netherlands, v. 151, p. 307-313, 2011.

KOOHMARAIE, M. et al. Interventions to reduce/eliminate *Escherichia coli* O157:H7 in groundbeef. **Meat Science,** Amsterdam, The Netherlands, n. 77, p. 90-96, 2007.

KOOHMARAIE, M. et al. Post-harvest interventions to reduce/eliminate pathogens in beef. **Meat Science,** Amsterdam, The Netherlands, n. 71, p. 79-91, 2005.

KORNACKI, J. L.; JOHNSON, J. L. *Enterobacteriaceae,* Coliforms, and *Escherichia coli* as Quality and safety indicators. In: AMERICAN PUBLIC HEALTH ASSOCIATION (APHA). **Compendium of Methods for the Microbiological Examination of Foods.** 4. ed. Washington, DC, 2001. chap. 8, p. 69-82.

KUNICKA, A. Evaluation of the TEMPO® SYSTEM: an automated method for food microbiological quality control. **Journal of Biotechnology,** Amsterdam, Netherlands, n. 131, p. 69-72, 2007.

LARA, G. H. B. et al. Occurrence of *Mycobacterium* spp. and other pathogens in lymph nodes of slaughtered swine and wild boars *(Sus scrofá).* **Research in veterinary Science,** Amsterdam, The Netherlands, n. 90, p. 185-188, 2011.

LAW, D. Virulence factors of *Escherichia coli* 0157 and other Shiga toxin-producing *E. coli.* **Journal of Applied Microbiology,** New Jersey, v. 88, p. 729-745, 2000.

LAWRIE, R. A. **Meat Science.** 6. ed. Porto Alegre: Artmed, 2005.

LAWSON, L. G. et al. Cost-effectiveness of *Salmonella* reduction in Danish abattoirs. **International Journal of Food Microbiology,** Amsterdam, The Netherlands, n. 134, p. 126-132, 2009.

LEE, G. Y. et al. Prevalence and classification of pathogenic *Escherichia coli* isolated from fresh beef, poltry and pork in Korea.

International Journal of Food Microbiology, Amsterdam, Netherlands, n. 134, p. 196-200, 2009.

LEOMIL, L. et al. Frequency of Shiga toxin-producing *Escherichia coli* (STEC) isolates among diarrheic and non-diarrheic calves in Brazil.
Veterinary Microbiology, Amsterdam, The Netherlands, v. 97, p. 103-109,2003.

LIMA, E. do S. C. et al. Isolation of *Salmonella* sp and *Staphylococcus aureus* in the pig slaughter process as a subsidy to the Hazard Analysis and Critical Control Points (HACCP) system. **Pesquisa Veterinária Brasileira,** Seropédica, v. 24, n. 4, p. 185-190, 2004.

LO FO WONG, D. M. A. et al. Epidemiology and control measures for *Salmonella* in pigs and pork. **Livestock Production Science,** Amsterdam, The Netherlands, v. 76, n. 3, p. 215-222, 2002.

LOMBARD, B.; LECLERCQ, A. Validation of innovative food microbiological methods according to the EN ISO 16140 Standard. **Food Analytical Methods,** Berlin, v. 4, n. 2, p. 163-172, 2010.

MANI-LÓPEZ, E.; GARCÍA, H. S.; LÓPEZ-MALO, A. Organic acids as antimicrobials to control *Salmonella* in meat and poultry products. **Food Research International,** Amsterdam, The Netherlands, v. 45, n. 2, p. 713- 721, Mar. 2012.

MATARAGAS, M.; SKANDAMIS, P. N.; DROSINOS, E. H. Risk profiles of pork and poultry meat and risk ratings of various pathogen/product combinations.
International Journal of Food Microbiology, Amsterdam, The Netherlands, n. 126, p. 1-12, 2008.

MCMEEKIN, T. A. et al. Ecophysiology of food-bome pathogens: essential knowledge to improve food safety. **International Journal of Food Microbiology,** Amsterdam, The Netherlands, n. 139, p. 64-78,2010.

MENG, J.; ZHAO, S.; DOYLE, M. Pathogenic *Escherichia coli*. In: AMERICAN PUBLIC HEALTH ASSOCIATION (APHA).
Compendium of Methods for the Microbiological Examination of Foods. 4. ed. Washington, DC, 2001.

MENG, J.; ZHAO, S.; DOYLE, M. Virulence genes of Shiga toxin- producing *Escherichia coli* isolated from food, animals and humans.
International Journal of Food Microbiology, Amsterdam, Netherlands, n. 45, p. 229-235, 1998.

MENG, J. et al. Enterohemorrhagic *Escherichia coli*. In: DOYLE, M. P.; BEUCHAT, L. R.; MONTVILLE, T. J. **Food microbiology:** fundamentals and frontiers. 3. ed. Washington, DC: ASM, 2007.

MENG, J. et al. Pathogenic *Escherichia coli.* In: AMERICAN PUBLIC HEALTH ASSOCIATION (APHA). **Compendium of Methods for the Microbiological Examination of Foods.** 4. ed. Washington, DC, 2001.

MIELE, M.; MACHADO, J. S. Panorama of Brazilian pork. **AgroAnalysis,** São Paulo, v. 30, n. 1, esp. Suinocultura p. 36-42, jan. 2010.

MOLDENHAUER, J. OverView of rapid microbiological methods. In: ZOUROB M.; ELWARY, S.; TURNER, A. P. F. **Principies of bacterial detection:** biosensors, recognition receptors and microsystems. Berlin: Springer, 2008.

MORA, A. et al. Antimicrobial resistance of Shiga toxin (verotoxin): producing *Escherichia coli* O157:H7 and non-O157 strains isolated from humans cattle, sheep and food in Spain. **Research in Microbiology,** Amsterdam, The Netherlands, v. 156, p. 793-806, 2005.

MORTON, R. D. Aerobic plate count. In: AMERICAN PUBLIC HEALTH ASSOCIATION (APHA). **Compendium of Methods for the Microbiological Examination of Foods.** 4. ed. Washington, DC, 2001. chap. 7, p. 63-67.

MURMANN, L.; SANTOS, M. C.; CARDOSO, M. Prevalence, genetic, characterisation and antimicrobial resistance of *Salmonella* isolated from fresh pork sausages. **Food Control,** Amsterdam, The Netherlands, n. 20, p. 191-195,2009.

NAKAO, H. et al. Subtyping of Shiga toxin 2 variants in human-derived Shiga toxin-producing *Escherichia coli* strains isolated in Japan. **FEMS Immunology and Medicai Microbiology,** New Jersey, n. 34, p. 289- 297, 2002.

NAKAZAWA, M.; AKIBA, M.; SAMESHIMA, T. Swine as a potential reservoir of Shiga-Toxin-producing *Escherichia coli* O157:H7 in Japan. **Emerging Infectious Diseases,** Atlanta, GA, v. 5, n. 6, nov./dec. 1999.

NESBAKKEN, T.; SKJERVE, E. Interruption of microbial cycles in farm animals from farm to table. **Meat Science,** Amsterdam, The Netherlands, v. 43, p. 47-57, 1996.

NESBAKKEN, T. et al. Occurrence of *Yersinia enterocolitica* and *Campylobacter* spp. in slaughter pigs and consequences for meat inspection, slaughtering, and dressing procedures. **International Journal of Food Microbiology,** Amsterdam, The Netherlands, n. 80, p. 231-240, 2003.

NISSEN, H.; MAUGESTEN, T.; LEA, P. Survival and growth of *Escherichia coli* O157:H7, *Yersinia enterocolitica* and *Salmonella enteritidis* on decontaminated and untreated meat. **Meat Science,** Amsterdam, The Netherlands, n. 57, p. 291-298, 2001.

O'LOUGHLIN, E. V.; ROBINS-BROWNE, R. M. Effect of Shiga toxin and Shiga-like toxins on eukaryotic cells. **Microbes and Infection,** Amsterdam, The Netherlands, v. 3, p. 493-507, 2001.

OJO, O. E. et al. Potentially zoonotic Shiga toxin-producing *Escherichia coli* serogroups in the faeces and meato f food-producing animals in Ibadan, Nigeria. **International Journal of Food Microbiology,** Amsterdam, The Netherlands, n. 142, p. 214-221, 2010.

OLIVEIRA, M. et al. Occurrence of *Salmonella* spp. in samples from pigs slaughtered for consumption: a comparison between ISO 6579:2002 and 23S rRNA Fluorescent in Situ Hybridisation method. **Food Research International,** Amsterdam, The Netherlands, v. 45, n. 2, p. 984- 988, 2010.

ORDONEZ, M. G. Ensaios farmacológicos in vitro para evaluar actividad antigiardsica. **Revista Cubana de Farmacia,** La Habana, Cuba, v. 35, n. 1, p. 66-73, 2001.

OTEIZA, J. M. et al. Isolation and characterisation of Shiga toxin- producing *Escherichia coli* from precooked sausages (morcillas).
International Journal of Food Microbiology, Amsterdam, Netherlands, v. 23, p. 283-288, 2006.

OTERO, A.; GARCIA-LOPÉZ, M. L.; MORENO, B. Rapid microbiological methods in meat and meat products. **Meat Science,** Amsterdam, The Netherlands, v. 49, n. suppl. I, p. 179-189, 1998.

OWEN, M.; WILLIS, C.; LAMPH, D. Evaluation of the TEMPO® most probable number technique for the enumeration of Enterobacteriaceae in food and dairy products. **Journal of Applied Microbiology,** New Jersey, v. 109, p. 1810-1816, 2010.

PANISELLO, P. J. et al. Application of foodbome disease outbreak data in the development and maintenance of HACCP systems. **International Journal of Food Microbiology,** Amsterdam, The Netherlands, n. 59, p. 221-234, 2000.

PARDI, M. C. **Memória da inspeção sanitária e industrial de produtos** de **origem animal no Brasil:** o Serviço de Inspeção Federal- SIF. Brasília: Columbia, 1996.

PEARCE, R. A. et al. Studies to determine the critical points in pork slaughter hazard analysis and critical control point systems.
International Journal of Food Microbiology, Amsterdam, Netherlands, n. 90, p.

331-339, 2004.

PIPEK, P. et al. Colour changes after carcasses decontamination by steam and lactic acid. **Meat Science,** Amsterdam, Netherlands, n. 69, p. 673-680, 2005.

PIPEK, P. et al. Decontamination of pork carcasses by steam and lactic acid. **Journal of Food Engineering,** Amsterdam, Netherlands, n. 74, p. 224- 231,2006.

PRATA, C. B. **Occurrence of *Escherichia coli* O157:H7 in cattle slaughtered in an establishment authorised for export in the city of Barretos-SP, Brazil.** 2009. 74f. Dissertation (Master's in Agricultural Microbiology)-Faculty of Agricultural and Veterinary Sciences, Universidade Estadual Paulista, Jaboticabal, 2009.

REAGAN, J. O. et al. Trimming and washing of beef carcasses as a method of improving the microbiological quality of meat. **Journal of Food Protection,** Des Moines, IA, v. 59, n. 7, p. 751-756, 1996.

REITER, M. G. et al. Comparative study of alternative methods for food safety control in poultry slaughterhouses. **Food Analytical Methods,** Berlin, v. 3, p. 253-260, 2010.

RIVAS, T.; VIZCAINO, J. A.; HERRERA, F. J. Microbial contamination of carcasses and equipment from an Iberian pig slaughterhouse. **Journal of Food Protection,** Des Moines, IA, v. 63, n. 12, p. 1670-1675,2000.

ROST, E. **Photos from personal archive.** São Miguel do Oeste, 2010.

ROTTA, A. **Photos from personal archive.** Chapecó, 2010.

SAIDE-ALBORNOZ, J. J. et al. Contamination of pork carcasses during slaughter, fabrication and chilled storage. **Journal of Food Protection,** Des Moines, IA, v. 58, n. 9, p. 993-997, 1995.

SANDRINI, C. N. M. et al. Verotoxigenic *Escherichia coli*: isolation and prevalence in 60 dairy cattle farms in the Pelotas region, RS Brazil. **Ciência Rural,** Santa Maria, v. 37, n. 1, p. 175-182, 2007.

SANT'ANA, A. S.; CONCEIÇÃO, C.; AZEREDO D. R. P. Comparison between the Simplater Tpc-Ci and Petrifilmr Ac rapid methods and conventional plate count methods for the enumeration of Mesophilic Aerobes in ice cream. **Ciência e Tecnologia de Alimentos,** Campinas, v. 22, n. 1, p. 60-64, jan./abr. 2002.

SARDINHA, L. et al. Decontamination of swine carcasses with chlorine. In: INTERNATIONAL CONGRESS OF MEAT SCIENCE AND TECHNOLOGIES, 47, 2001, Krakovia, Poland. **Proceedings...** Krakovia, Poland, 2001.

SARIMEHMETOGLU, B. et al. Detection of *Escherichia coli* O157:H7 in ground beef using immunomagnetic separation and multiplex PCR. **Food Control,** Amsterdam, The Netherlands, v. 20, p. 357-361, 2009.

SCHWARZ, P. et al. *Salmonella enterica'.* isolation and seroprevalence in pigs slaughtered in Rio Grande do Sul. **Arquivo Brasileiro de Medicina Veterinária e Zootecnia,** Belo Horizonte, v. 61, n. 5, p. 1028-1034, 2009.

SCIENTIFIC Opinion on a quantitative microbiological risk assessment of *Salmonella* in slaughter and breeder pigs. **European Food Safety Authority,** Parma, Italy, v. 8, n. 4, p. 1-90, 2010.

SCOTT, H. et al. Swine health impact on carcass contamination and human foodbome risk. **Public health Reports,** Washington, DC, v. 123, may/june 2008.

SEIXAS, F. N.; TOCHETTO, R.; FERRAZ, S. M. Presence of *Salmonella* sp. in pork carcasses sampled at different points of the processing line. **Ciência Animal Brasileira,** Goiânia, v. 10, n. 2, p. 634-640, Apr./Jun. 2009.

SILBERNAGEL, K.; LINDBERG, K. G. 3M™ Petrifilm™ Enterobacteriaceae count plate method for enumeration of enterobacteriaceae in selected foods: collaborative Study. **Journal of**

AO AC International, Gaithersburg, MD, v. 86, n. 4, 2003. Available at: <http://www.3m.com/intl/kr/microbiology/regu info/3.pdf>. Accessed on: 05 May 2012.

SILBERNAGEL, K.M; LINDBERG, K. G. Evaluation of the 3M Petrifilm Enterobacteriaceae count plate method for the enumeration of Enterobacteriaceae in foods. **Journal of Food Protection,** Des Moines, IA, v. 65, n. 9, p. 1452-1456, sept. 2002.

SILVA, A. S. et al. *Escherichia coli* strains from oedema disease: o serogroups and genes for Shiga toxin, enterotoxins and F18 fimbriae. **Veterinary Microbiology,** Amsterdam, The Netherlands, v. 80, p. 227-233,2001.

SILVA, M. C. et al. Prevalence of *Salmonella* sp. in pigs slaughtered in the state of Mato Grosso. **Ciência Rural,** Santa Maria, v. 39, n. 1, p. 266-268, jan./feb. 2009.

SILVA, M. P.; CAVALLI, D. R.; OLIVEIRA, T. C. R. M. Evaluation of the coliform standard at 45°C and comparison of the efficiency of the multiple tube and Petrifilm EC techniques in detecting total *coliforms* and *Escherichia coli* in food. **Ciência e Tecnologia de Alimentos,** Campinas, v. 26, n. 2, p. 352-359, Apr./Jun. 2006. Available at: <http://www.scielo.br/pdf/cta/v26n2/30183 .pdf>. Accessed on: 03 June 2012.

SILVA, N. et al. Occurrence of *Escherichia coli* O157:H7 in meat products and sensitivity of detection methods. **Ciência e Tecnologia de Alimentos,** Campinas, v. 21, n. 2, p. 223-227, May/Aug. 2001.

SILVA, N. et al. Occurrence of *Escherichia coli* 0157 in vegetables and resistance to vegetable disinfection agents. **Ciência e Tecnologia de Alimentos,** Campinas, v. 23, n. 2, p. 167-173, May/Aug. 2003.

SILVEIRA, J. B. **Investigation of *Escherichia coli* O157:H7 in ground beef in the state of Rio Grande do Sul, Brazil.** 2010. 46 f.
Dissertation (Master's Degree in Food Science and Technology)- Federal University of Rio Grande do Sul, Porto Alegre, 2010.

SMIGIC, N. et al. Intracellular pH as an indicator of viability and resuscitation of *Campylobacter jejum* after decontamination with lactic

acid. **International Journal of Food Microbiology,** Amsterdam, The Netherlands, n. 135, p. 136-143, 2009.

SOBESTIANSKY, J. et al. **Clínica e patologia suína.** 2. ed. Goiânia: Art3, 1999.

SODERGARD, A.; STOLD, M. Properties of acid lactic based polymers and their correlation with the composition. **Progress in Materials Science,** Amsterdam, The Netherlands, v. 27, n. 6, p. 1123-1163,2002.

SOFOS, J. N. Challenges to meat safety in the 21st[st] century. **Meat Science,** Amsterdam, Netherlands, n. 78, p. 3-13, 2008.

SOFOS, J. N.; BELK, K. E.; SMITH, G. C. Nonacid meat decontamination technologies: model studies and commercial applications. **International Journal of Food Microbiology,** Amsterdam, The Netherlands, n. 44, p. 171-188, 1998.

SOFOS, J. N.; BELK, K. E.; SMITH, G. C. Processes to reduce contamination with pathogenic microorganisms in meat.
INTERNATIONAL CONGRESS OF MEAT SCIENCE AND TECHNOLOGIES, 45, 1999, Yokohama, Japan. **Proceedings...** Yokohama, Japan, 1999. p. 596-605.

SOFOS J. N.; GEORNARAS I. OverView of current meat hygiene and safety risks and summary of recent studies on biofilms, and control of *Escherichia coli* O157:H7 in nonintact, and *Listeria monocytogenes* in ready-to-eat, meat products. **Meat Science,** Amsterdam, The Netherlands, n. 86, p. 2-14, 2010.

SOHIER D. et al. TEMPO® system: AFNOR validation according to the ISO 16140 Standard 2006. **Inside Laboratory Management,** Marcy l'Etoile, France, may/june 2007.

SOHIER, D.; RANNOU, M.; GORSE, F. A. Développement, expert lab at the AFNOR validation committee, France, BioMérieux, France. TEMPO® system :AFNOR validation according to the ISO 16140 Standard. 2006. Available at: <http://www.biomerieux-usa.com/upload/ISO%2016140%20validation%20by%20AFNOR%20on%20TVC,%20TC%20and%20EC,%20Adria-2.pdf>. Accessed on: 12 May 2011.

SORRIBES, C. H. Rapid methods and automation in food microbiology. **Alimentaria:** Revista de Investigación, Tecnologia y Seguridad, Madrid, p. 109-113, Apr. 2008.

SPRICIGO, D. A. et al. Prevalence, quantification and antimicrobial resistance of *Salmonella* serovars isolated from fresh pork sausage. **Ciência e Tecnologia de Alimentos,** Campinas, v. 28, n. 4, p. 779-785, Oct./Dec. 2008.

STOPFORTH, J. D. et al. Influence of organic acid concentration on survival of *Listeria monocytogenes* and *Escherichia coli* O157:H7 in beef carcass wash water and on model equipment surfaces. **International Journal of Food Microbiology,** Amsterdam, Netherlands, n. 20, p. 651-660, 2003.

SWANENBURG, M. et al. *Salmonella* in slaughter pigs: the effect of logistic slaughter procedures of pigs on the prevalence of *Salmonella* in pork. **International Journal of Food Microbiology,** Amsterdam, The Netherlands, n. 70, p. 231-242, 2001.

TORLAK, E.; AKAN, I. M.; GOKMEN, M. Comparison of TEMPO® EC and TBX medium for the enumeration of *Escherichia coli* in cheese. **Letters in Applied Microbiology,** New Jersey, n. 47, p. 566-570,2008.

VAN NETTEN, P.; MOSSEL, D. A.; HUIS in't VELD, J. H. J. Microbial changes on freshly slaughtered pork carcasses due to "hot" lactic acid decontamination. **Food Safety,** Ashtown, Dublin, n. 17, p. 89-111, 1997.

VAN NETTEN, P.; MOSSEL, D. A; HUIS in't VELD, J. H. J. Lactic acid decontamination of fresh pork carcasses: a pilot plant study. **International Journal of Food Microbiology,** Amsterdam, Netherlands, n. 25, p. 1-9, 1995.

VARELA-HERNÁNDEZ, J. J. et al. Isolation and characterisation of Shiga toxin-producing *Escherichia coli* 0157:H7 from beef carcasses at a slaughter plant in Mexico. **International Journal of Food Microbiology,** Amsterdam, The Netherlands, v. 113, p. 237-241, 2007.

VASCONCELOS, E. C. et al. The microbiota of carcasses and sheep meat treated with acetic acid, vacuum-packed and matured for 48 days. **Ciência e Tecnologia de Alimentos,** Campinas, v. 22, n. 3, p. 272-277, 2002.

VELOSO, G. M. Microbiology of meat: part I. In: GILL, J. I. **Manual de inspeção sanitária de carnes.** 2. ed. Lisbon: Calouste Gulbenkian Foundation, 2000. v. 1.

VELUSAMY, V. et al. An overview of foodbome pathogen detection: in the perspective of biosensors. **Biotechnology Advances,** Amsterdam, Netherlands, n. 28, p. 232-254, 2010.

VERNOZY-ROZAND, C. et al. Prevalence of *Escherichia coli* O157:H7 in industrial minced beef. **Letters in Applied Microbiology,** New Jersey, v. 35, p. 7-11, 2002.

VERSETTI, D. et al. **Evaluation of the tempo® system for total viable count, total coliforms and e. coli enumeration in meat products.** Durham: bioMérieux, 2011. Available at: <http://www.biomerieuxusa.com/upload/TVC%20TC%20and%20EC%20vs%20ISO%20in%20meat%2Q(Inalca)-1 .pdf>. Accessed on: 12 May 2011.

VIANNA, A. T. **Os Suínos:** criação prática e económica. 4. ed. São Paulo: Nobel, 1974.

VIEIRA-PINTO, M. et al. Relationship between tonsils and mandibular lymph nodes conceming *Salmonella* sp. infection. **Food Research International,** Amsterdam, The Netherlands, v. 45, n. 2, p. 863-866, 2010.

WANI, S. A. et al. Subtype analysis of *stXJ stX2* and *eae* genes in Shiga toxin-producing *Escherichia coli* (STEC) and typical and atypical enteropathogenic *E. coli* (EPEC) from lambs in india. **The Veterinary Journal,** Amsterdam, **The** Netherlands, n. 182, p. 489-490, 2009.

WONG, T. L.; MACDIARMID, S.; COOK, R. *Salmonella, Escherichia coli* O157:H7 and *E. coli* biotype 1 in a pilot survey of imported and new Zealand pig meats. **International Journal of Food Microbiology,** Amsterdam, The Netherlands, n. 26, p. 177-182, 2009.

Printed by Books on Demand GmbH, Norderstedt / Germany